NANOTECHNOLOGY ENVIRONMENTAL HEALTH AND SAFETY

NANOTECHNOLOGY ENVIRONMENTAL HEALTH AND SAFETY

Edited by:

Abhishek Gupta

ARCLER PRESS

www.arclerpress.com

Nanotechnology Environmental Health and Safety

Abhishek Gupta

Arcler Press
2010 Winston Park Drive,
2nd Floor
Oakville, ON L6H 5R7
Canada
www.arclerpress.com
Tel: 001-289-291-7705
001-905-616-2116
Fax: 001-289-291-7601
Email: orders@arclereducation.com

ISBN: 978-1-77407-335-3 (Hardcover)

DECLARATION

Some content or chapters in this book are open access copyright free published research work, which is published under Creative Commons License and are indicated with the citation. We are thankful to the publishers and authors of the content and chapters as without them this book wouldn't have been possible.

ABOUT THE EDITOR

Dr. Abhishek Gupta is currently a National Post-Doctoral Fellow (NPDF) at All India Institute of Medical Sciences (AIIMS), New Delhi. He has completed their Doctoral Degree (Ph.D) from School of Basic Sciences, Indian Institute of Technology (IIT) Mandi HP India. He obtained his B.Sc. in Biological Sciences from MJP R University Bareilly India and M.Sc. Biotechnology from School of Biotechnology, Devi Ahilya University, Indore India. His recent research is focused on carbon nanoparticles (C-dots) for toxic metal and biomolecule sensing system. He has more than 15 publications in reputed international journal (Nano Letter, Chem Comm, JPC Letter, ACS Applied Materials and Interfaces, Nature Scientific Reports and etc.) with average Impact factor > 5.50.

TABLE OF CONTENTS

LIST OF CONTRIBUTORS

Dr. Abhishek Gupta

Mohamed El-Helaly
Community Medicine Department, Faculty of Medicine, Mansoura University, Egypt
Director EHOHS, King Abdulaziz Medical City, Riyadh, Saudi Arabia

Anderson DS
Center for Environmental Health Sciences, Biomedical and Pharmaceutical Sciences, University of Montana, Missoula, MT, USA

Sydor MJ
Center for Environmental Health Sciences, Biomedical and Pharmaceutical Sciences, University of Montana, Missoula, MT, USA

Fletcher P
Center for Environmental Health Sciences, Biomedical and Pharmaceutical Sciences, University of Montana, Missoula, MT, USA

Holian A
Center for Environmental Health Sciences, Biomedical and Pharmaceutical Sciences, University of Montana, Missoula, MT, USA

Charles W. Schmidt

In-Ju Kim
Department of Industrial Engineering and Engineering Management, College of Engineering, University of Sharjah, Sharjah, United Arab Emirates

Paul A. Schulte
National Institute for Occupational Safety and Health, Centers for Disease Control and Prevention, Cincinnati, Ohio, USA

Fabio Salamanca-Buentello
University of Toronto Joint Centre for Bioethics and Canadian Program on Genomics and Global Health, Toronto, Ontario, Canada

Amela Groso
Occupational Safety and Health, School of Basic Sciences, Ecole Polytechnique Fédérale de Lausanne Switzerland

Alke Petri-Fink
Advanced Particles Group, Department of Chemistry, University of Fribourg, Switzerland

Arnaud Magrez
Laboratory of Nanostructures and Novel Electronic Materials, Ecole Polytechnique Fédérale de Lausanne, Switzerland

Michael Riediker
Institute for Work and Health (Institut universitaire romand de Santé au Travail), Lausanne, Switzerland

Thierry Meyer
Occupational Safety and Health, School of Basic Sciences, Ecole Polytechnique Fédérale de Lausanne Switzerland

Eileen D. Kuempel
National Institute for Occupational Safety and Health (NIOSH), Education and Information Division, Nanotechnology Research Center, Cincinnati, OH 45226, USA

Charles L. Geraci
National Institute for Occupational Safety and Health (NIOSH), Education and Information Division, Nanotechnology Research Center, Cincinnati, OH 45226, USA

Paul A. Schulte
National Institute for Occupational Safety and Health (NIOSH), Education and Information Division, Nanotechnology Research Center, Cincinnati, OH 45226, USA

Duane Hammond
Division of Applied Research and Technology, National Institute for Occupational Safety and Health, Centers for Disease Control and Prevention, 4676 Columbia Parkway, Mail Stop R5, Cincinnati, OH 45226, USA

Alberto Garcia
Division of Applied Research and Technology, National Institute for Occupational Safety and Health, Centers for Disease Control and Prevention, 4676 Columbia Parkway, Mail Stop R5, Cincinnati, OH 45226, USA

H. Amy Feng
Division of Applied Research and Technology, National Institute for Occupational Safety and Health, Centers for Disease Control and Prevention, 4676 Columbia Parkway, Mail Stop R5, Cincinnati, OH 45226, USA

Bernd Giese
University of Bremen, Faculty of Production Engineering, Department of Technology Design and Technology Development, Badgasteiner Str, 1 28359, Bremen, Germany
University of Natural Resources and Life Sciences, Institute of Safety and Risk Sciences, Borkowskigasse 4, 1190, Vienna, Austria

Fred Klaessig
Pennsylvania Bio Nano Systems, Doylestown, Pennsylvania, 18901, United States
Center for Environmental Implications of Nanotechnology (UC CEIN), University of California Santa Barbara, Santa Barbara, California, 93106-5131, United States

Barry Park
GBP Consulting Ltd, Purton, UK

Ralf Kaegi
Eawag, Swiss Federal Institute of Aquatic Science and Technology, Überlandstrasse 133, 8600, Dübendorf, Switzerland

Michael Steinfeldt
University of Bremen, Faculty of Production Engineering, Department of Technology Design and Technology Development, Badgasteiner Str, 1 28359, Bremen, Germany

Henning Wigger
University of Bremen, Faculty of Production Engineering, Department of Technology Design and Technology Development, Badgasteiner Str, 1 28359, Bremen, Germany
Empa, Swiss Federal Laboratories for Materials Science and Technology, Lerchenfeldstrasse 5, CH-9014, St. Gallen, Switzerland

Arnim von Gleich
University of Bremen, Faculty of Production Engineering, Department of Technology Design and Technology Development, Badgasteiner Str, 1 28359, Bremen, Germany

Fadri Gottschalk
ETSS AG, Engineering, technical and scientifc services, CH7558, Strada, Switzerland

Thomas AJ Kuhlbusch
Air Quality & Sustainable Nanotechnology, Institute of Energy and Environmental Technology e.V. (IUTA), D-47229 Duisburg, Germany
Center for Nanointegration Duisburg-Essen (CeNIDE), University Duisburg-Essen, D-47057, Germany

Christof Asbach
Air Quality & Sustainable Nanotechnology, Institute of Energy and Environmental Technology e.V. (IUTA), D-47229 Duisburg, Germany

Heinz Fissan
Air Quality & Sustainable Nanotechnology, Institute of Energy and Environmental Technology e.V. (IUTA), D-47229 Duisburg, Germany
Center for Nanointegration Duisburg-Essen (CeNIDE), University Duisburg-Essen, D-47057, Germany

Daniel Göhler
Research Group Mechanical Process Engineering, Institute of Process Engineering and Environmental Technology, Faculty of Mechanical Engineering, Technische Universität Dresden (TUD), D-01062 Dresden, Germany

Michael Stintz
Research Group Mechanical Process Engineering, Institute of Process Engineering and Environmental Technology, Faculty of Mechanical Engineering, Technische Universität Dresden (TUD), D-01062 Dresden, Germany

Vivek K. Bajpai
Department of Energy and Materials Engineering, Dongguk University-Seoul, 30 Pildong-ro 1-gil, Seoul, 04620, South Korea

Madhu Kamle
Department of Forestry, North Eastern Regional Institute of Science and Technology, Nirjuli, 791109, Arunachal Pradesh, India

Shruti Shukla
Department of Energy and Materials Engineering, Dongguk University-Seoul, 30 Pildong-ro 1-gil, Seoul, 04620, South Korea

Dipendra Kumar Mahato
Department of Agriculture and Food Engineering, Indian Institute of Technology Kharagpur, West Bengal, 721302, India

Pranjal Chandra
Department of Biosciences and Bioengineering, Indian Institute of Technology Guwahati, Guwahati, 781039, Assam, India

Seung Kyu Hwang
Department of Biological Engineering, Biohybrid Systems Research Center (BSRC), Inha University, 100 Inha-ro, Nam-gu, Incheon, 22212, South Korea

Pradeep Kumar
Department of Forestry, North Eastern Regional Institute of Science and Technology, Nirjuli, 791109, Arunachal Pradesh, India

Yun Suk Huh
Department of Biological Engineering, Biohybrid Systems Research Center (BSRC), Inha University, 100 Inha-ro, Nam-gu, Incheon, 22212, South Korea

Young-Kyu Han
Department of Energy and Materials Engineering, Dongguk University-Seoul, 30 Pildong-ro 1-gil, Seoul, 04620, South Korea

LIST OF ABBREVIATIONS

APS	Aerodynamic Particle Sizer
AUC	analytical ultracentrifugation
CNF	carbon nanofibers
CNT	carbon nanotubes
CPC	Condensation Particle Counter
CSH	calcium silicate hydrate
DI	Dustiness Index
DT	Dust Track
EDX	energy dispersive X-ray spectroscopy
EEPS	Engine Exhaust Particle Sizer
ELPI	electrical low-pressure impactor
EM	electron microscopy
ENM	engineered nanomaterial
ENP	engineered nanoparticle
ESP	electrostatic precipitator
FMPS	Fast Mobility Particle Sizer
HHCPC	hand-held condensation particle counter
HHPC	hand-held particle counter
ICBA	International Carbon Black Association
ICP-MS	Inductively Coupled Plasma - Mass Spectrometry
M	mass
MC	mass concentration
MOUDI	micro-orifice uniform deposit impactor
NC	number concentration
NRP	number of released particles
NSAM	Nanoparticle Surface Area Monitor

OPC	Optical Particle Counter
PM	particulate matter
PSD	particle size distribution
SA	Surface Area
SEM	scanning electron microscope
SIMS	secondary-ion mass spectroscopy
SMPS	Scanning Mobility Particle Sizer
TEM	transmission electron microscope
TP	thermophoretic precipitator
UNPA	universal nanoparticle analyzer
V	volume
WRASS	Wide Range Aerosol Sampling System
XPS	X-ray photoelectron spectroscopy

PREFACE

The book provides the detailed overview about the environment health and safety, where researcher exposed to the nanotechnology work. A variety of reaction process can generate a toxicated environment for the bad impression on human health. Importantly workplace and handling an important parameter to secure and protect from the chemically developed nanoparticles exposure. To ensure worker safety, and uncertain nanotechnology EHS risks, health surveillance strategies should be considered in order to identify and track health problems attributable to workplace operations. Toxic chemicals, their fumes and carcinogenic concern must be a point of care during lab establishment and complete protocol need to be follow during the lab instrument installation. Quantifying employee health before the employee begins a job function provides a baseline health profile to which future health screens may be compared. In this regard, the coupling of baseline and periodic health screens may help identify potential health hazards at their earliest and most correctable stages. Employee must be received a complete demonstration of lab-based safety during joining the particular research active area and must be provided a confidence for independent work. Nanomaterial can be vaporize in many processes, so must be a full aeration and fume hood based facility are the key term for prevention and care. Toxicity level optimization must be functionalized before the reaction processing in any sensitity level. This edited book provided specific section and details for accurate safety and concern about the human health working in nanomaterials-based research in any private or govt. aided organization

SUMMARY

Environmental emissions (i.e., air, wastewater, solid wastes) from nanotechnology facilities may contain engineered nanomaterials. It is currently unclear what environmental risks are posed by such emissions. Proactive approaches to evaluating the properties of these emissions and/or managing them may be important for protecting public health and the natural environment. In the US, environmental regulatory policies are in place that applies to products and emissions containing engineered nanomaterials.

Key Points:

While initial safety concerns focus primarily on potential human health hazards in the workplace, it is important for nanotechnology facility operators to consider downstream implications of emerging nanotechnologies on the natural environment. Facility managers may consider mapping their manufacturing processes and/or laboratory handling procedures to identify potential release scenarios in air emissions, process water, and/or solid waste streams. For organizations with sufficient resources, efforts may be taken to modify processing steps and/or implement control technologies to reduce or eliminate unintended environmental emissions.

Many nanotechnology organizations are unclear as to how current state and federal environmental regulations relate specifically to nanotechnology. Studies have shown that many existing environmental statutes (the Toxic Substances Control Act, the Resource Conservation and Recovery Act, the Clean Air Act, and the Federal Insecticide, Fungicide, and Rodenticide Act) apply to nanomanufacturing and associated products and/or wastes. Other environmental statutes may apply or may soon apply to nanomanufacturing. These include the Comprehensive Environmental Response, Compensation and Liability Act, the Clean Water Act, and other new approaches (Environmental Management Systems/Innovative Regulatory Approaches) customized specifically for nanomanufacturing facilities.

Comprehensive life cycle assessment approaches (i.e., cradle to grave) may help identify, mitigate, and communicate possible environmental hazards associated with engineered nanomaterials and nano-enabled products.

Introduction

QUANTITATIVE ANALYSIS AND SAFETY ISSUES OF NANOTECHNOLOGY IN HEALTHCARE RESEARCH

It will be difficult to achieve a smooth translation for nanotechnology if universal quantitative analytical techniques have not been fully implemented in the characterization process.

In the past decade, nanotechnology has made the headlines on numerous occasions. Interestingly, it seems those nano-sized materials have garnered far more attention than their tiny size would seem to justify in healthcare research, including approaches for personalized medicine. At the same time, however, precautious issues have also been raised as well, due to the implication of nanoscale physics on environment, health and safety for producers to end users.

What, then, makes the development of nanotechnology such a controversial issue? No doubt, nanomaterials certainly offer promises. Yet health experts are concerned that this area of research has grown so fast that without appropriate regulation, unintended consequences will outweighs the potential economic and health benefits. Today, with more than

---

Citation: Wellington Pham, 2012, Quantitative Analysis and Safety Issues of Nanotechnology in Healthcare Research, Open Access Int. J Mol Biomark Diagn, 2012 3:e111. doi:10.4172/2155-9929.1000e111.

1,000 nanoscale materials available commercially (Nanotechnology and Environmental, Health, and Safety: Issues for Consideration. Congressional Research Service Report 2011), it is vital to protect the public from harm.

Quite obviously, there are lessons to be learned from previous instances of pharmaceutical development fiascos, such as the case of Thalidomide. But while Thalidomide was produced by a single company; today's nanomaterials are being manufactured by a variety of firms and laboratories worldwide. Indeed, a single component could be contributed by any of a number of global raw materials suppliers, thus making identification of the source of a problem and pinpointing the responsible party a much lengthier and difficult process than one might believe.

In healthcare research, nanoscale particles represent a burgeoning area of research. This technique has the potential to offer fundamental changes in diagnostics and novel interventions for disease treatment. Naturally, those nanomaterials are nonspecific for biological study. If specific ligands are incorporated in the nanoparticles, then the result would be a highly specific and targeted probe. In fact, all these advances in nanotechnology are possible only because the inherent multivalency of nanoparticles offers myriad surface modifications for molecular imaging, drug delivery, vaccine delay release applications and more. However, the process of colloidal surface modification and characterization are devious. For instance, we must consider the large body of work from relevant literature which describes that nanomaterials having been mixed with the ligands in the presence of an activating reagent, such as dicyclohexylcarbodiimide or a water soluble version of 1-ethyl-3-(3-dimethylaminopropyl)carbodiimide, and assuming the products are present after dialysis. In reality, the process of chemical coupling under aqueous conditions is insufficient. Intense characterization via analytical chemistry is necessary not only to determine if the desired products have been obtained, but also quantification of the labeled products is crucial and should be used as a parameter to ensure reproducibility and safety.

It is not easy to overcome current practices since quantification of nanomaterials is expensive. After all, not every laboratory is equipped with the sophisticated analytical instruments needed to perform the work. Further, researchers prefer quick turnovers of their work through publications. Still, such considerations cannot be used as excuses.

Characterization of the chemical product is not a matter of if, but rather when. Yet regardless of the answer, one thing is certain: if nanotechnology is being groomed for human application, then appropriate characterization of the chemistry at the beginning of the conceptual processes may prove beneficial overtime, particularly with regard to GMP and IND processes. After all, in the face of a decade spent validating a spectrum of cell and animal models, the failure of the materials to meet FDA requirements because of lack of quantitative analysis would be intolerable.

Several scientific panels have already begun deliberating the safety issues of nanotechnology in healthcare. Among the important topics of discussion is the need to safeguard of human health via strenuous implementation of analytical techniques to characterize nanoproducts. The lack of such an approach would hinder the translation of this promising science from bench to bedside. With that in mind, each of us has a major role to play shares the responsibility to emphasize this important issue at all levels. Remaining passive or even silent about this challenge would only restrict the contributions nanotechnology could make in biomedical research. Clearly, the need for quantitative analysis as a forerunner to the impeding translation process cannot be marginalized.

IMPACT : A different view of the environment

APPLYING A METHOD COMMONLY USED IN MICROBIOLOGY PROVIDES A NEW WAY TO STUDY THE INTERACTION OF NANOPARTICLES WITH ENVIRONMENTAL SAMPLES

What is the potential hazard of nanomaterials to the environment? How many nanoparticles can be released into the environment before they start causing problems? These are basic questions, but answering them is necessary to create the appropriate regulations for the utilization of engineered — hence not naturally occurring — nanomaterials in products available to society.

Though the questions may be basic, answering them is not simple. The properties of nanoparticles, including their potential environmental effects, depend on many factors including the exact size, composition and shape. Eventually, the environmental effect of nanoparticles depends also on the specific conditions considered, and it is therefore encouraging that studies of environmental nanotechnology under realistic conditions are increasing.

One such study is included in this issue. Jacob Metch and co-workers report the results of their study of the interaction of gold nanoparticles with a microbial community extracted from wastewater sludge using metagenomic analysis, discovering that after a period of several weeks the shape of nanoparticles has a clear effect on the microbial community examined. In a nutshell, metagenomic analysis involves the high-throughput sequencing of DNA extracted from an environmental sample, and aims at capturing the diversity of a whole natural microbial community, as opposed to single types of microbial organism as typically studied using cloned cultures in the lab.

Though metagenomics is a relatively common technique in environmental microbiology, its use to study the effects of nanoparticles is a new approach and can provide information based on a large number of data that would not be possible to obtain otherwise. Metch and colleagues were especially interested in exploring the effect of metal nanoparticles coated with different substances and with different morphologies. Gold was chosen primarily due to its relative inertness, so that the effect of two types of coating and two shapes — spheres and rods — could be studied independently from the potential effects of the metal.

The particles were released in the sludge sample for a period of several weeks and metagenomic analysis was performed to evaluate both taxonomic and functional gene expression. The results reveal an effect of both coating and morphology, but perhaps surprisingly the effects of differences in morphology were more evident.

The work shows that by using this type of high-throughput analysis it is possible to explore general trends using, in principle, a large variety of nanomaterials. In cases like that of Metch and co-workers, the analysis can reveal unexpected effects that will be fascinating to investigate in detail. Furthermore, the fact that the analysis works by default on actual environmental samples is a particularly attractive aspect.

Perhaps indirectly, the potential effect of the nanoparticle morphology is also a corroboration of the general feeling within the environmental nanotechnology community that more specific information should be provided on nanoparticles, whether they are used in products and can be potentially released in the environment, or they are simply used in research studies reported in the scientific literature. The particle size, size distribution and shape, and eventually the methods used to determine all these specifications are but a few pieces of information that would contribute to the development of a more complete picture, improving transparency and reproducibility of results.

For scientific literature, publishers can facilitate transparency by encouraging authors to provide relevant details. Along these lines, for a few years Nature journals including *Nature Nanotechnology* have requested authors to fill in a checklist for life sciences experiments, detailing information such as the sample size or the number and types of replica. The document is now used as a reporting summary and is published alongside the corresponding paper. A similar document was also created for manuscripts reporting solar cells, and since 1 March 2018 a reporting summary on behavioural and social sciences has been introduced. Work is in progress on a reporting summary in ecology, evolution and environmental studies. This will be partly relevant for manuscripts in environmental nanotechnology as it will include details about the environmental sample chosen. However, at this stage the inclusion of details on the characterization of the materials involved is not being contemplated. We are fully aware that these details are more significant for publications on environmental effects of nanotechnology, and we are intending to collaborate with the scientific community to establish the best way to report essential details.

1

Nanotechnology, Occupational Health and Safety Concerns

Mohamed El-Helaly[1,2]

[1]Community Medicine Department, Faculty of Medicine, Mansoura University, Egypt
[2]Director EHOHS, King Abdulaziz Medical City, Riyadh, Saudi Arabia

Nanotechnology is the manipulation of matter on a near-atomic scale to produce new structures, materials and devices. Nanoparticles (NPs), the building blocks of nanotechnology, are the objects with at least one dimension smaller than 100 nanometer [1,2]. NPs find numerous applications in many fields, starting with electronics, throughout medicine, cosmetology, and ending with automotive and construction industries [3]. Nanotechnology, nanomedicine and nanotoxicology are complementary disciplines aimed at the improvement of human life. Nanomedicine will develop applications for novel and superior diagnostic, therapeutic and preventive measures. Nanotoxicity provides for the necessary safety assessment of nano-enabled products [4,5]. Exciting achievements based on nanotechnology and nanomedicine await us in the future; yet there are many challenges to get it right and recognize and avoid potential risks associated with these new developments where nanotoxicology will have a crucial role [6].

Citation: El-Helaly M (2013) Nanotechnology, Occupational Health and Safety Concerns. Occup Med Health Aff 1:116. doi: 10.4172/2329-6879.1000116.

Concerns have been expressed about risks posed by engineered nanomaterials (ENMs), their potential to cause undesirable effects, contaminate the environment and adversely affect susceptible parts of the population [7,8]. Toxicity of NPs depends on many factors, for example: size, shape, chemical composition, solubility, surface area and surface charge. Risk assessment related to human health, should be integrated at all stages of the life cycle of the nanotechnology, starting at the point of conception and including research and development, manufacturing, distribution, use and disposal or recycling [3]. Fundamentally, risk assessment involves an estimation of the potential for exposure and characterization of hazard. Potential routes of NPs exposure include inhalation, dermal, oral, and in the case of biomedical applications, parenteral. Toxicity resulting from NPs exposure could occur at the various portals of entry, such as the lungs and skin, or at distant sites. Exposure to nanomaterials could occur during their development, manufacture, use, or following disposal [9].

NPs translocation and uptake by the body occurs after inhalation exposure (neuronal uptake, translocation across lung epithelium, and ingestion), oral exposure (ingestion), and dermal exposure depending on the characteristics of the NPs under investigation [9,10]. With the exception of airborne particles delivered to the lung, information on the biological fate of NPs including distribution, accumulation, metabolism, and organ specific toxicity is still minimal [2,11].

Nanosizing of some particulates, increases pathologic and physiologic responses, including inflammation, fibrosis, allergic responses, genotoxicity, and carcinogenicity, and may alter cardiovascular and lymphatic function [10,12-15]. Knowing how the size and physiochemical properties of NPs affect bioactivity is important in assuring that the exciting new products of nanotechnology are used safely [11,16].

A tragic case in point appears to be a recent report about worker exposure to heated polystyrene fumes and polyacrylate NPs in an unventilated confined space for several months, resulting in progressive pulmonary fibrosis, pleural effusions and granuloma formation with fatal outcome [17]. Regardless as to whether the NPs had caused the severe pathology – which is unclear, based on the information provided – holding on to extremely poor industrial hygiene conditions at the workplace was completely irresponsible. Preventing exposure is key, and that can readily be achieved today with appropriate engineering technology and personal protection equipment [6].

Even without being able to perform a quantitative risk assessment for ENMs, due to the lack of sufficient data on exposure, biokinetics and organ toxicity, until we know better it should be made mandatory to prevent exposure by appropriate precautionary measures/regulations and practicing best industrial hygiene to avoid future horror scenarios from environmental or occupational exposures. Similarly, safety assessment for medical applications as key contribution of nanotoxicology to nanomedicine relies heavily on nano-specific toxicological concepts and findings and on a multidisciplinary collaborative approach involving material scientists, physicians and toxicologists [6].

Occupational safety and health (OSH) concerns are receiving considerable attention in nanoscience and nanotechnology research and development. Policymakers and others have urged that research on nanotechnology's EHS implications be developed alongside scientific research in the nanotechnology domain rather than subsequent to applications [18]. Occupational physicians would thus be required to keep abreast and update themselves on toxicological and health and safety developments in this growing industry. There is also the need to look beyond the factory fence to consider safety and environmental impact of NPs containing products at all stages of the life cycle [8,10].

In conclusion, the current knowledge of OSH in nanotechnology includes the following: (i) NPs can be measured using standard measurement methods (respirable mass or number concentration), (ii) workplace exposures to NPs can be reduced using engineering controls and personal protective equipment, and (iii) current toxicity testing and risk assessment methods are applicable to NPs. Yet, to ensure protection of workers' health, research is still needed to develop (i) sensitive and quantitative measures of workers' exposure to NPs, (ii) validation methods for exposure controls, and (iii) standardized criteria to categorize hazard data, including better prediction of chronic effects [2].

REFERENCES

1. Becker H, Herzberg F, Schulte A, Kolossa-Gehring M (2011) The carcinogenic potential of nanomaterials, their release from products and options for regulating them. Int J Hyg Environ Health 214: 231-238.
2. Bolt HM, Marchan R, Hengstler JG (2012) Recent developments in nanotoxicology. Arch Toxicol 86: 1629-1635
3. Furukawa F, Doi Y, Suguro M, Morita O, Kuwahara H, et al. (2011) Lack of skin carcinogenicity of topically applied titanium dioxide nanoparticles in the mouse. Food Chem Toxicol 49: 744-749.
4. Hubbs AF, Mercer RR, Benkovic SA, Harkema J, Sriram K, et al. (2011) Nanotoxicology--a pathologist's perspective. Toxicol Pathol 39: 301-324.
5. Kuempel ED, Geraci CL, Schulte PA (2012) Risk assessment and risk management of nanomaterials in the workplace: translating research to practice. Ann Occup Hyg 56: 491-505.
6. Kuhlbusch TA, Asbach C, Fissan H, Göhler D, Stintz M (2011) Nanoparticle exposure at nanotechnology workplaces: a review. Part Fibre Toxicol 8: 22.
7. Magaye R, Zhao J, Bowman L, Ding M (2012) Genotoxicity and carcinogenicity of cobalt-, nickel- and copper-based nanoparticles. Exp Ther Med 4: 551-561.
8. Mitka M (2012) Committee calls for framework to assess the safety of nanotechnology materials. JAMA 307: 1123-1124.
9. Nyström AM, Fadeel B (2012) Safety assessment of nanomaterials: implications for nanomedicine. J Control Release 161: 403-408.
10. Oberdörster G (2010) Safety assessment for nanotechnology and nanomedicine: concepts of nanotoxicology. J Intern Med 267: 89-105.
11. Seaton A, Tran L, Aitken R, Donaldson K (2010) Nanoparticles, human health hazard and regulation. J R Soc Interface 7: S119-129.
12. Sng J, Chia SE (2011) Nanotechnology health and safety--what can occupational health professionals do? Ind Health 49: 545-547.
13. SnopczyÅ„ski T, GÃ³ralczyk K, Czaja K, StruciÅ„ski P, Hernik A, et al. (2009) [Nanotechnology--possibilities and hazards]. Rocz Panstw Zakl Hig 60: 101-111.

14. Song Y, Li X, Du X (2009) Exposure to nanoparticles is related to pleural effusion, pulmonary fibrosis and granuloma. Eur Respir J 34: 559-567.
15. Stern ST, McNeil SE (2008) Nanotechnology safety concerns revisited. Toxicol Sci 101: 4-21.
16. Tsuda H, Xu J, Sakai Y, Futakuchi M, Fukamachi K (2009) Toxicology of engineered nanomaterials - a review of carcinogenic potential. Asian Pac J Cancer Prev 10: 975-980.
17. Ventola CL (2012) The nanomedicine revolution: part 2: current and future clinical applications. P T 37: 582-591.
18. Youtie J, Porter A, Shapira P, Tang L, Benn T (2011) The use of environmental, health and safety research in nanotechnology research. J Nanosci Nanotechnol 11: 158-166.

2

Nanotechnology: The Risks and Benefits for Medical Diagnosis and Treatment

Anderson DS, Sydor MJ, Fletcher P and Holian A

Center for Environmental Health Sciences, Biomedical and Pharmaceutical Sciences, University of Montana, Missoula, MT, USA

Nanotechnology is being actively developed for many applications in the medical field, including drug delivery, biosensors and medical imaging. These nanomaterials are being advanced as novel and more targeted treatments for difficult to manage diseases such as cancers. Other materials are being developed as alternatives to conventional antibiotics in treating infections. The use of engineered nanomaterials (nanoparticles) offer the ability to transport therapeutics to specific sites of a disease, thus reducing the off target toxicity of many drugs. This is especially true in the use of chemotherapeutics where off target reactions cause serious side effects in cancer patients. Additionally, the field of medical imaging can be improved with the ability for the specific targeting of diseased tissues at resolutions not capable with current technologies.

Citation: Anderson DS, Sydor MJ, Fletcher P, Holian A (2016) Nanotechnology: The Risks and Benefits for Medical Diagnosis and Treatment. J Nanomed Nanotechnol 7: e143. doi:10.4172/2157-7439.1000e143.

As mentioned above, one of the primary uses of nanotechnology in the medical field has been the area of targeted drug delivery. It is crucial to deliver a drug to a desired target site in a controlled manner while not causing additional adverse health effects to the patient. To date the principle type of particle used in drug delivery systems is composed of lipids or polymers chosen for their biocompatibility. These systems have been used to deliver therapeutics more targeted or efficiently. For instance, nanospheres conjugated with disease specific antibodies or peptides can result in greatly increased local doses of treatments to sites of disease while avoiding high systemic levels of the therapeutic [1]. Biocompatible nanospheres have also been used to transport therapeutics with poor bioavailability. For instance, biologic treatments such as insulin and calcitonin, that cannot be delivered by conventional methods as an oral treatment, have successfully been packaged in hollow nanoparticles that protect it from degradation in the gastrointestinal tract allowing for systemic delivery of the drug and avoiding alternative methods of delivery such as subcutaneous injection [2]. However, there has also been a trend to use nanoparticles in medicine that have been produced from materials that are regarded as less biocompatible than those discussed above.

Nanoparticles have become a significant interest as a drug delivery system due to their small size and large surface area. The small sizes of nanoparticles increase efficacy for accurate intracellular uptake of the drug in the desired cellular targets and for accurate biodistribution [3,4]. The large surface area makes it easy to manipulate the particle into carrying high levels of drug or other compounds with ease [5]. Another important factor is the structural stability of nanoparticles to effectively deliver the drug over a long period of time without degradation occurring before it reaches the cellular target. In breast cancer treatment, Sabzichi, et al. used nanostructured lipid carriers to carry melatonin in human breast cancer cells which inhibited tumor proliferation and induced apoptosis. The study results indicated that the nano-carriers had effective biocompatibility and low cytotoxicity [5].

In contrast to the beneficial outcomes, using nanoparticles for drug delivery raises various safety concerns. The small size is beneficial, but it could have negative effects. This is a particularly important point as new and more durable materials are used in the production of these nanoparticles. Some nanoparticles can cause inflammation and fibrosis as a result of causing phagolysosomal membrane permeability, formation of reactive oxygen species and activation of the NLRP3 inflammasome [6]. The small size, in turn, indicates a large surface area which could be harmful by exposing more surface molecules to cellular components. Another concern is the preparation and stabilization processes of the nanoparticles for drug delivery since chemical reducing agents and radiation can exacerbate cytotoxicity. One antimicrobial testing study addressed this concern. Pradeepa, et al. used an environmentally friendly technique by means of bacterial exopolysaccharide to synthesize gold nanoparticles. They found that this new technique did not interfere with the antibiotic-coated gold nanoparticles ability to inhibit Gram negative and Gram positive bacteria activity [7]. Many things are still uncertain when using nanoparticles as a drug delivery system, which is why more research needs to be performed at relevant doses.

Nanoparticles have been proposed for improved systems in medical imaging for disease diagnosis. Much of the potential for nanomaterials as diagnostic agents comes from their ability to enhance contrast in spectroscopy. In particular, superparamagnetic iron oxide has been shown to enhance magnetic resonance imaging and as a result can aid in the detection of liver metastases [8]. Angiogenesis is a key hallmark of cancer and thus its detection would be of importance. Nanoparticles can be used for targeting sites of angiogenesis and enhancing diagnostic imaging. For example, cyanoacrylate microbubbles can be conjugated to ligands specific to biomarkers, such as vascular endothelial growth factor receptor and αvβ3integrin, which are more abundant with increased angiogenesis [9]. This conjugation allows for ultra sound detection of tumor phenotypes and quantification of these biomarkers, as well as, indicating responses to treatment.

Even though there are various applications for nanomaterials as diagnostic agents, some considerations need to be made with regard to their pharmacokinetic properties. Diagnostic agents must possess high target site specificity, low toxicity, and rapid clearance for the fraction unbound to the target. The size of the nanomaterial is also an important consideration. Agents larger than 5 nm will have much slower renal clearance than smaller agents. On the other hand, materials larger than 1 μm will not undergo renal clearance and can, as a result, have high target specificity with a smaller unbound fraction needing to be cleared [10].

With the number and combination of materials used and the many different shapes and structures in which nanoparticles are being produced increasing each year there is a vital need to determine the properties of nanoparticles that may produce adverse reactions in patients receiving these materials as part of diagnosis or treatment. For instance, quantum dots have been shown to have properties that could improve medical imaging in the diagnosis of cancer [11]. However, quantum dots can contain cadmium, selenium, arsenic, and lead. While these particles have many advantages in diagnostics and imaging due to the unique properties that make them stand out from the biological background, these toxic metals are of concern [12]. Another example is the use of silver nanoparticles as an antibiotic to treat infection of multiple-drug resistant bacteria. Ionized silver from the silver nanoparticles has been found to be persistent in the organs of animals exposed and the long term effects of this persistent silver is poorly understood [13].

While nanotechnology, including the medical use of nanoparticles, hold great promise to improve the diagnosis and treatment of many diseases, we must not lose sight of the necessity to thoroughly test the nanomaterials so that they do not create unexpected adverse effects. This will require a balance between the safety of the materials used and the efficacy of the treatment. Different considerations of toxicology should be used when proposing that a given nanoparticle be used for a treatment of stage four cancer as compared to one to be used for a routine diagnostic medical imaging. Therefore, a situational approach should be used when assessing the benefits and drawbacks to using nanoparticles in medical diagnosis and treatment.

REFERENCES

1. Sonaje K, Lin KJ, Wey SP, Lin CK, Yeh TH, et al. (2010) Biodistribution, pharmacodynamics and pharmacokinetics of insulin analogues in a rat model: Oral delivery using pH-responsive nanoparticles vs. subcutaneous injection. Biomaterials 31: 6849-58.
2. Pridgen E, Alexis F, Farokhzad OC (2015) Polymeric Nanoparticle Drug Delivery Technologies for Oral Delivery Applications. Expert opinion on drug delivery 12: 1459-1473.
3. Bao H, Zhang Q, Xu H, Yan Z (2016) Effects of nanoparticle size on antitumor activity of 10-hydroxycamptothecin-conjugated gold nanoparticles: in vitro and in vivo studies. Int J Nanomedicine. 11: 929-40
4. Semete B, Booysen L, Lemmer Y, Kalombo L, Katata L, et al. (2010) In vivo evaluation of the biodistribution and safety of PLGA nanoparticles as drug delivery systems. Nanomedicine 6: 662-71.
5. Sabzichi M, Samadi N, Mohammadian J, Hamishehkar H, Akbarzadeh M, et al. (2016) Sustained release of melatonin: A novel approach in elevating efficacy of tamoxifen in breast cancer treatment. Colloids Surf B Biointerfaces 145: 64-71.
6. Hamilton RF, Buford M, Xiang C, Wu N, Holian A (2012) NLRP3 inflammasome activation in murine alveolar macrophages and related lung pathology is associated with MWCNT nickel contamination. Inhalation Toxicology 24: 995-1008.
7. Pradeepa, Vidya SM, Mutalik S, UdayaBhat K, Huilgol P, et al. (2016) Preparation of gold nanoparticles by novel bacterial exopolysaccharide for antibiotic delivery. Life Sci 15: 171-179.
8. Rappeport ED, Loft A, Berthelsen AK, von der Recke P, Larsen PN, et al. (2007) Contrast-enhanced FDG-PET/CT vs. SPIO-enhanced MRI vs. FDG-PET vs. CT in patients with liver metastases from colorectal cancer: a prospective study with intraoperative confirmation. ActaRadiol 48: 369-78.
9. Palmowski M, Huppert J, Ladewig G, Hauff P, Reinhardt M, et al. (2008) Molecular profiling of angiogenesis with targeted ultrasound imaging: early assessment of antiangiogenic therapy effects. Mol Cancer Ther 7: 101-9.
10. Baetke SC, Lammers T, Kiessling F (2015) Applications of nanoparticles for diagnosis and therapy of cancer. Br J Radiol 88: 20150207.

11. Santra S, Dutta D, Walter GA, Moudgil BM (2005) Fluorescent Nanoparticle Probes for Cancer Imaging. Technology in Cancer Research and Treatment 4: 593-602.
12. Pericleous P, Gazouli M, Lyberopoulou A, Rizos S, Nikiteas N, et al. (2012) Quantum dots hold promise for early cancer imaging and detection. International Journal of Cancer 131: 519-528.
13. Lee JH, Kim YS, Song KS, Ryu HR, Sung JH, et al. (2013) Biopersistence of silver nanoparticles in tissues from Sprague-Dawley rats. Particle and Fibre Toxicology 10: 36-36.

3

Nanotechnology-Related Environment, Health, and Safety Research: Examining the National Strategy

Charles W. Schmidt

Pick up a tube of sunscreen, a tennis racquet, an iPod, or any number of other consumer products, and there's a good chance that it's been "nano-enabled," meaning it contains nanoscale particles designed to give it some beneficial feature. An estimated $147 billion worth of nano-enabled commercial and consumer products were sold in 2007, according to Lux Research, a market analysis firm in New York City. Citing the firm's latest estimates, Lux analyst David Hwang predicts that figure could top $3.1 trillion by 2015, reinforcing a broad view that nanotechnology is fueling a new industrial revolution.

Yet nanotechnology's spread through the market has been met with mounting concerns over the potential human health effects of these miraculous materials. Because of their small size—100 nano-meters or less—nanomaterials have unique physical properties that can influence their uptake, distribution, and behavior in the body. Indeed, some nano-particles have been shown to penetrate into cells, where they can trigger inflammatory responses and oxidative stress.

Citation: Charles W. Schmidt, 2009, Nanotechnology-Related Environment, Health, and Safety Research: Examining the National Strategy, Open Access Environ Health Perspect. 2009 Apr; 117(4): A158–A161. 2009, https://doi.org/10.1289/ehp.117-a158

Canada and California recently took the unprecedented step of imposing mandated disclosure requirements on nanomaterial use and toxicity assessment. Issued 29 January 2009, Canada's law targets domestic companies and institutions that manufacture or buy more than 1 kilogram of nanomaterial per year. According to the new regulations, these entities must now reveal how much nanomaterial they use, how they use it, and what they know about its toxicity. California's law, issued 2 February 2009, limits its scope to carbon nanotubes, a class of nanomaterial used in electronics, optics, and biomedical applications. Under the new regulation, by February 2010 companies that manufacture, import, or export carbon nanotubes in California must disclose information about the toxicity and environmental impacts of their products.

Meanwhile, experts in nanotoxicology and risk assessment have become increasingly polarized, represented on one side by the National Research Council (NRC) and on the other by the National Nanotechnology Initiative (NNI), a government-wide collaboration coordinated by the National Science and Technology Council in the Executive Office of the President. In February 2008, the Nanotechnology Environmental and Health Implications (NEHI) Working Group of the NNI released a document titled *Strategy for Nanotechnology-Related Environmental, Health, and Safety Research*. This document is meant to present the U.S. government's agenda for studying nano-particle hazards, and describes 246 related projects that were ongoing in 2006, representing a combined investment for that year of $68 million. The document also purports to "address prioritized research areas . . . and to advance knowledge and support risk decision-making—both of which are essential for the responsible development of nanotechnology."

Clayton Teague directs the National Nano-technology Coordination Office, which was responsible for drafting the federal strategy. He says the strategy was developed in extensive consultation with regulatory agencies, research organizations, the business community, and nongovernmental organizations. "We believe the strategy represents needs and agreements about what the agencies plan to do," he says. "Funding agencies are telling us that they're using the document to formulate solicitations for future research in this area."

But on 25 February 2009, a panel assembled by the NRC issued its own report, describing what it calls serious shortcomings in the strategy document. According to the NRC panel, which was assembled at the request of the NNI, the strategy exposes weaknesses in the government's understanding of potential nanotechnology risks today and does not adequately address how they will be assessed in the future. NRC panel member Mark Weisner, a professor of civil and environmental engineering at Duke University, claims that many of the research programs described in the NNI's document don't actually address environmental, health, and safety (EHS) concerns. "If you take this portfolio at face value, it overstates the true level of effort in federally financed [nano-technology-related] EHS research," he says.

Toxicity Unknowns

No case of human toxicity has been linked to the roughly 2,000 types of nano-materials in commercial use or development today. Yet those risks can't be ruled out, says Günter Oberdörster, a professor of environmental medicine at the University of Rochester School of Medicine and Dentistry. According to Oberdörster, multiwalled carbon nanotubes have been found to elicit responses similar to those seen with fibers of chrysotile asbestos, a known human carcinogen. Oberdörster emphasizes these findings have been seen only in rodents given carbon nanotubes at extremely high doses by injection. What's needed now, he says, are toxicity data generated by inhalation routes that mimic real-life human exposure. Oberdörster says he and others in the field are currently working on such studies.

Predictions about nanotechnology risk have emerged from inhalation research, specifically studies targeting ultrafine soot particles with nanoscale dimensions. Upon inhalation, some of these particles traverse epithelial and endothelial cells to reach the blood and lymph circulation, which carries them to potentially sensitive sites, including the bone marrow, lymph nodes, spleen, heart, and central nervous system. *In vitro* and animal studies show these particles can—depending on the dose and chemical composition—induce a range of inflammatory effects, whereas epidemiologic findings link them to respiratory and cardiovascular diseases.

All nanoparticles have high surface-to-mass ratios, which makes them uniquely reactive in the body. "Chemical reactions tend to occur at particle surfaces," explains Jeff Morris, associate director for science in the U.S. Environmental Protection Agency (EPA) Office of Science Policy. "Given that their surface area exceeds their mass, nano-particles tend to be more reactive than larger particles with the same chemical makeup."

Engineered nanoparticles and soot differ in key ways, however. In particular, soot is heterogeneous in terms of particle size, chemistry, surface characteristics, and other constituents, whereas engineered nano-particles—within product categories—have uniformly identical shapes, including spheres, tubes, wires, rings, and planes. Given their similar high surface-to-volume ratios, both types of particles could trigger comparable biologic effects, Oberdörster adds.

But particle uniformity might also influence the kinetics and toxicity of nano-materials in unknown ways. For instance, Andrew Maynard, science advisor to The Project on Emerging Nanotechnologies, a collaboration of the Pew Charitable Trust and the Woodrow Wilson Center for International Scholars, proposes that some of the particles in soot and other heterogeneous mixtures could be more harmful than others. "In that case, the toxicity of the harmful particles is diluted by the presence of others that are less so," he explains. "But when you engineer particles with precise characteristics, you lose that dilution factor, and the chance of producing something uniformly dangerous increases."

LOOKING FOR A STRATEGY

The NRC panelists would like to see a national, health-based strategy for nanotech-nology research, with defined goals, milestones, and mechanisms for assessing progress. Maynard stresses the need isn't just to ensure the safety of nano-enabled products, but also to avert a public backlash against the technology, which could grow if health risks aren't seen to be adequately addressed. Yet the NNI strategy document—NRC panelists claim—is simply a compendium of federally funded projects without any unifying vision or sense of shared purpose. Each of the projects listed by the NNI is grouped under one of five research categories: instrumentation, metrology, and analytical methods; nanomaterials and human health; nanomaterials and the environment; human and environmental exposure assessment; and risk management methods.

In Maynard's view, these projects aren't adequately organized around questions of public concern. Instead, they reflect investigator-motivated studies, piqued by each scientist's own interests, he asserts. "Scientists don't like being told what to do," Maynard acknowledges. "But there's a disconnect between what might interest them and what companies and regulators who deal with nanotechnology actually need."

Sally Tinkle, senior science advisor in the NIEHS Office of the Director and cochair of the Nanoscale Science, Engineering, and Technology (NSET) subcommittee of the National Science and Technology Council, says federal agencies have had to make do without any federal appropriation specifically for nanotoxicology research. "The agencies have to fund what they can under flat budgets with competing research priorities," she says.

A prepublication copy of the NRC report was leaked to the press 10 December 2008. The ensuing media attention prompted the NNI on 5 January 2009 to post an 18-page rebuttal on its website, http://www.nano.gov/, which stated that the strategy document is not, and was never intended to be, a strategic plan or an implementation plan, but rather a higher-level description of the interagency approach to nanotechnology-related EHS research; "[i]t was written as a strategy document for federal agencies in order to coordinate, encourage cooperation, and where possible to implement collaborative research actitivies." The rebuttal goes on to list what it calls technical errors in the NRC assessment. Although such errors were corrected in the final February 2009 release, the NRC did not change its overall conclusions.

Of paramount importance, Wiesner says, is that exposure and toxicity research in nanotechnology be balanced appropriately. "We don't want the toxicity work to get too far out in front of the exposure work, and yet the toxicity work tells us where we should focus our exposure studies," he says. "There's a delicate balance we need to achieve here, and that's something the whole research community is struggling with right now."

Exposure research in nanotechnology does come with unique challenges, Wiesner acknowledges. Scientists have yet to develop widely accepted methods for introducing nanomaterials into living systems such as cell cultures, for instance. As nanomaterial surfaces interact with cell macromolecules and salts, their properties can change in mysterious ways. And those transformations, Wiesner says, directly influence interpretations of effective exposure and dose response.

QUESTIONS REGARDING TRANSPARENCY

Meanwhile, given mounting scrutiny, industry has become more sensitive about its public image vis-à-vis nanomaterials, according to Ellen Kenney, a senior research analyst with the Bethesda investment firm Calvert Group, Ltd. Companies that don't use nano-materials have begun stating so in their shareholder reports, she says, in a reflection of how these materials might be viewed as a public liability.

Indeed, many companies are reluctant to reveal their nanomaterial use and toxicity data voluntarily. With its Nanoscale Materials Stewardship Program (NMSP), launched 28 January 2008, the EPA urges companies to report available information about the engineered nanoscale materials they manufacture, import, process, or use. This voluntary two-year effort is intended to help inform eventual regulatory decisions about nanomaterials. According to Lux Research, the total number of companies engaged in nanotech production or use could reach 1,000. The EPA reached out to more than 150 companies and 11 trade associations, Hwang says, but by the time the NMSP published its interim report on 12 January 2009, only 29 companies had responded. In total, these companies disclosed data on 123 nanomaterial compounds.

Jim Willis, who directs the EPA Chemical Control Division, says those results left agency personnel with mixed feelings. "On the one hand, we thought it was pretty good responsiveness for a volunteer program," he says. "On the other, we know there are hundreds of other nanomaterials that weren't reported. And that indicates clearly that we need to do more if we want to get a better handle on what's being produced, at what levels, and how humans are being exposed."

Sources interviewed for this article unanimously agreed that nanomaterials promise valuable benefits for society, among them better drugs; stronger, lighter products; and better environmental and energy technologies. But nanoparticle toxicity data need to be made more widely available to ensure public support for the technology. Jennifer Sass, a staff scientist with the Natural Resources Defense Council, says such data typically wind up in company reports instead of in publicly available, peer-reviewed research journals. And Julia Moore, deputy director of The Project on Emerging Nanotechnology, claims the public has limited access to information about which companies use nanomaterials and how. "That information isn't in the hands of government regulators," she says. "It's in the hands of market analysts on Wall Street, and they're not going to let it go without a price."

Those on the industry side believe many interest groups have significantly overhyped the dangers of nanomaterials. "Fear-mongering both inhibits industry efforts to encourage companies utilizing nanotechnology to do so in a visible way by 'branding' it and makes it more difficult for entrepreneurs to raise capital and find partners to bring new innovations to market," says Sean Murdoch, executive director of the Nanobusiness Alliance, a trade group based in Skokie, Illinois. He cites the comparison between carbon nanotubes and chrysotile asbestos as an example. He points out that, unlike asbestos, which for decades was mined at million-ton quantities by unprotected workers, carbon nanotubes are processed in laboratories subject to strict safety protocols. Moreover, he says, the nanotubes themselves, once incorporated into products, have no bioavailability. “There’s no way the exposure scenarios are comparable,” he says.

But this argument does not address product end-of-life concerns, say many experts. Tinkle says, “There is still concern over exposure to nanoparticles at the end of the products’ life cycles, even if companies design the product to be completely safe for the immediate user. Once [a nano-enabled item] is thrown out and begins to decompose or degrade—or it begins to break down from day-to-day use—the particles can be released into the environment. Care needs to be taken to control the exposure throughout the product life cycle.”

For his part, Oberdörster suggests most nanoparticles may turn out to be benign under real-life exposure conditions. “I think there’s a certain amount of hype surrounding the toxicity issues,” he says. “However, until we know better, we should be careful and avoid exposure. You can do a lot of *in vitro* testing at high doses and identify a hazard, but you need the necessary exposure for a risk to be present.”

Still, assuming the growth trends continue, nanomaterials will be produced at ever-increasing quantities, and public and environmental exposures will rise commensurately. Given that reality, the industry’s future may well rest on its transparency to public scrutiny.

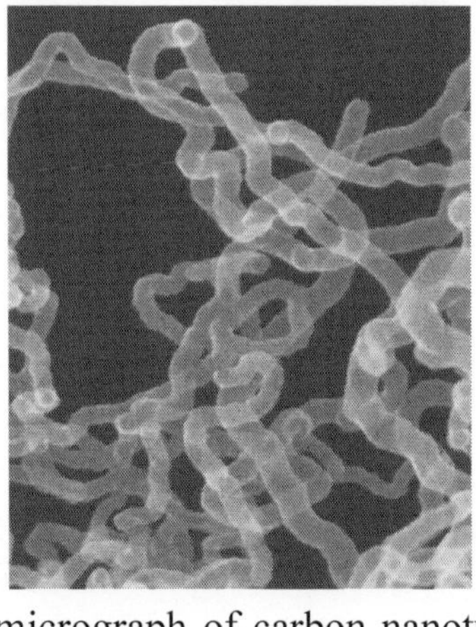

Scanning electron micrograph of carbon nanotubes, magnified 40,000 times.

The NEHI Working Group developed this . . . research strategy to accelerate progress in research to protect public health and the environment, and to fill gaps in, and—with the growing level of effort worldwide—to avoid unnecessary duplication of, such research.

— NNI Strategy Document

The process of composing the government's 2008 NNI document provided a unique and useful opportunity for coordination, planning, and consensus-building among NEHI-member federal agencies. . . . However, [the document] does not have the essential elements of a research strategy—it does not present a vision, contain a clear set of goals, have a plan of action for how the goals are to be achieved, or describe mechanisms to review and evaluate funded research and assess whether progress has been achieved in the context of what we know about the potential EHS risks posed by nanotechnology.

—NRC report

4

ERGONOMIC CHALLENGES FOR NANOTECHNOLOGY SAFETY AND HEALTH PRACTICES

In-Ju Kim

Department of Industrial Engineering and Engineering Management, College of Engineering, University of Sharjah, Sharjah, United Arab Emirates

INTRODUCTION

Nanotechnology has been broadly introduced to a wide range of industry fields such as aeronautics, agriculture, architectural design, bio-medical engineering, communication sciences, constructions, environmental science, food production, and information technology over the last decades [1-4]. Nanotechnology has the perspective to radically advance the efficiency of current industries and industrial products and considerable effects on the development of new products in all areas, ranging from disease diagnosis and treatment to environmental remediation. Now, nanotechnologies and nanomaterials are familiar terminologies and commonly applied to the field of industry designs. Although nanotechnology-based and nanomaterial-based products are generally accepted at the precompetitive stage, an increasing

Citation: In-Ju Kim, 2018, Risks, Ergonomic Challenges for Nanotechnology Safety and Health Practices, Open Access J Ergonomics 2016, 6:5, DOI: 10.4172/2165-7556.1000e160.

number of products and materials are becoming commercially available. These include nanoscale powders, solutions, and suspensions of nanoscale materials as well as composite materials and devices having nanostructures [5]. Nanoscale products and materials are also increasingly applied to a number of areas such as optoelectronic, electronic, magnetic, medical imaging, drug delivery, cosmetic, catalytic, and materials applications. Nanomaterials oien show unique physical and chemical properties that impart specific characteristics to making engineered materials. But, little is known about what effects from such exclusive features may have on people's safety and health issues. He current literature has reported that the physicochemical characteristics of nanomaterials can influence their effects on biological systems [6]. These attributes include chemical properties, the degree of agglomeration, oxidant generation potential, particle size, shape, surface area, and solubility [7]. Until the results from research studies can fully clarify the characteristics of nanoparticles that may pose health and safety risks, precautionary measures are warranted. Because of the wide range of possible applications of nanotechnology, continued assessment of the potential health and safety risks associated with exposure to nanomaterials is essential to safeguard their secure handling [8]. There is also an insufficient emphasis on application of nanomaterials in the domain of design ergonomics [9]. Therefore, ergonomists should make an effort to aware on applications of nanomaterials in the field of design ergonomics and safety and health implications of nanotechnologies in their operations.

NANOTECHNOLOGY SAFETY AND ERGONOMIC CONCERNS

Nanotechnology and its concerns

Nanotechnology is broadly defined by size. It mainly deals with particles of size less than 100 nm at least in one dimension [9,10]. These tiny-sized substances are known as nanomaterials and could be either natural or anthropogenic in their origins [9]. Nanotechnology involves the manipulation of matter on nanometer scales and offers the potential for unparalleled improvements in many different areas [11]. The capability to operate matters at the atomic or molecular level makes it possible to form new materials, structures, and devices that develop exclusive physical and chemical properties related to nanoscale structures. Developments in the nanotechnology have been progressed quicker than comprehending on the conceivable significances of their side effects. Thus, nanotechnology raises many of the same issues as any new technology, including concerns about the toxicity and environmental impact of nanomaterials [12]. Workers within the nanotechnology-related industries have the potential to be exposed to exclusively engineered materials with nano-scale sizes and shapes and physicochemical properties. Minimal information is currently available on dominant exposure routes, potential exposure levels, and material toxicity of nanomaterials [13]. This fact indicates that our knowledge on the nanomaterial is not well understood and capability to prevent hazards from the nanotechnology is limited as well. Because of extremely small sizes and characteristics of the nanomaterial, the nanoparticles could translocate to the brain or other organs of the employee. Hence, effectual solutions for the exposure control should comprise a right hazard measurement of the pretentious work areas.

Safety and Health Issues of Nanotechnology

Nanotechnologies and nanomaterials are rapidly emerging in many fields [9,14]. With large increases in demands, implications of nanomaterials on human health-related and safety-related issues need to be reviewed [15]. The recent literature reports that nanomaterials and nanoparticles have direct adverse health effects such as neuronal toxicity [16] and cardiopulmonary diseases [17]. Before applying nanomaterials, therefore, we should have thorough knowledge on nanoparticles-induced hazards and on how to handle them safely [15,16,18]. Because of their unique physicochemical properties, in many cases, it is difficult to assess toxic impacts of nanomaterials on biological systems [9]. However, workers may be constantly exposed to nanoparticles in different phases of working processes starting from manufacturing to packaging [9]. For instance, nanoparticles can enter into the human body through different routes [9] such as inhalation [19], ingestion [20], dermal route [21], and parenteral route [22]. Hence, prerequisite actions are necessary to create healthy and safe working environments. These arrangements should include supplying ventilation systems with fume hoods and high-efficiency air filtration, personal protecting equipment, and product stewardship policies that workers and managers can understand [13]. With such direct actions, identification of potential risks, development of novel approaches for risk assessment, and management to protect workers from any recognized adverse health and safety effects also should be preceded to protect workers working with nanomaterials. By implementing these solutions into safety and health management programs, the nanotechnology industry can accompany such inspiring new technologies to the marketplace whilst guard the workers, publics, and environments. The safety of nanotechnology has been continuously tested. There has been some as yet unresolved debate on the potential toxicity of a specific type of nanomaterial which has been associated with tissue damage in animal studies [23]. However, the majority of available data indicate that there is nothing uniquely toxic about nanoparticles as a class of materials. Whether actual or perceived, the potential health and safety risks associated with the manufacture and use of nanomaterials must be carefully studied in order to advance our understanding of this field of science and to realize the significant benefits that nanotechnology has to offer society, such as for cancer research, diagnostics, and therapy [23].

Ergonomic Affairs for Nanotechnology

Ergonomics is a multidisciplinary science which deals with designing or arranging workplaces, products, equip or devices, and systems so that it can fit the people or workers, their movements and cognitive functions in relation to their work performance [9,24-28]. Ergonomics is actually a design-driven discipline so that improvements in all the domains of nanotechnologies and nanomaterials seem to be possible by the application of ergonomic design principles. Nanotechnology may produce remarkable new types of materials with unique properties that can be harnessed by developing new concepts of support, adjustment, aesthetics, and comfort [9]. A recent study stated that nanotechnological revolution would profoundly affect not only our daily lives but also the science and practice of human factors/ergonomics [29]. Thus, nanotechnology and nanomaterials seem to be beneficial in the creation of both sustainable and ergonomic designs. Moreover, workers and/or people who interact with nanomaterials and nanoparticles may expose to potential health and safety risks. This seems to be an emerging issue which can be an opportunity for ergonomists to take over those typical challenges because ergonomics, due to its interdisciplinary nature, can play a vital role to solve these difficulties associated with nanotechnological development [9]. To learn more about managing nanotechnology exposures in the workplace, as well as other important controls as handling or working with nanotechnologies and nanomaterials, ergonomic approaches would be effective to improve safety and health issues in the nanotechnology industry. Specifically, it can be expected that ergonomic applications and trials in the field of nanotechnology with special emphasis on occupational safety and health would be one of the future directions of research to ergonomists and ergonomic society.

CONCLUSION

The emergence of nanotechnology and nanomaterials is generally considered to be a major technological revolution in many fields [30]. As a result, nanotechnologies and nano-related techniques have been applied to a range of industries such as aerospace, agriculture, car manufacturing, communication and information engineering and technology, construction, environmental science, healthcare, medicine, sports, and textile industries [9]. Utilization of devices and/or gears of extremely small dimensions (nanoscale) seem to appear as a new challenge for the man–machine compatibility. Thus, ergonomic interventions become recognizable solutions for designing optimal interfaces in the nanotechnology [9]. To date, however, there have only been limited studies on risk management practices, neither in manufacturing nor in the handling of nanomaterials within industrial units [30]. Although most of the important operations of nanomaterials in diverse technology and industry fields are well known today, applications of nanotechnology in the field of design with the incorporation of ergonomic principles are less discussed [9]. Future research needs to explore new ideas on how to apply the knowledge of nanotechnology in the field of design with appropriate consideration of ergonomic values. For example, increasing work efficiency, performance, and productivity by regulating temperature and reducing cognitive impedance, ensuring occupational safety, and providing health hazards measures using nanomaterials may require in-depth investigations. Therefore, it can be expected that ergonomists take this opportunity to apply diverse ergonomic applications and trials in the field of nanotechnology and nanomaterials with special emphasis on safety and health to develop a safer and healthier society and workplace.

REFERENCES

1. Michael FA, Paulo JF, Daniel LS (2009) Nanomaterials and Nanotechnologies: An Overview. Nanomaterials, In: Nanotechnologies and design: an introduction for engineers and architects. ButterworthHeinemann 1: 1-16.
2. Arivalagan K, Ravichandran S, Rangasamy K, Karthikeyan E (2011) Nanomaterials and its potential applications. Int J ChemTech Re 3: 534-538.
3. Mehta MD (2011) Understanding advances in nanotechnology: minimizing risks and maximizing benefits with application of the appropriate governance framework. Int J Nanotechnol Mol Comp 3: 1-11.
4. Mousavi SR, Rezaei M (2011) Nanotechnology in agriculture and food production. J Appl Enviro. Biol Sci 1: 414-419.
5. Eastlake A, Hodson L, Geraci C, Crawford C (2012) A critical evaluation of material safety data sheets (MSDSs) for engineered nanomaterials. Chem Health Saf 19: 1-8.
6. Navya, PN, Daima HK (2016) Rational engineering of physicochemical properties of nanomaterials for biomedical applications with nanotoxicological perspectives. Nano Convergence 3:1.
7. National Institute for Occupational Safety and Health (2009) Approaches to safe nanotechnology managing the health and safety concerns associated with engineered nanomaterials. NIOSH 2009-125.
8. National Institute of Environmental Health Sciences (2016). Nanomaterials.
9. Chowdhury A, Sanjog J, Reddy SM, Karmakar S (2012) Nanomaterials in the field of design ergonomics: present status. Ergon 55: 1453-1462.
10. National Nanotechnology Initiative (2016) What is Nanotechnology.
11. Elnashaie S, Danafar F, Hashemipour H (2015) From Nanotechnology to Nanoengineering, In: Nanotechnology for Chemical Engineers. Springer Science Business Media Singapore. Chapter 2: 79-178.
12. Buzea C, Pacheco II, Robbie K (2007) Nanomaterials and nanoparticles: Sources and toxicity. Biointerphases 2: MR17-MR71.
13. National Institute for Occupational Safety and Health (2016) Nanotechnology. Workplace Safety and Health Topics, He National Institute for Occupational Safety and Health (NIOSH). 14. Caruthers

SD, Winkline SA, Lanza GM (2007) Nanotechnological applications in medicine. Curr Opin Biol 18: 26-30.

14. Sequeira R, Genaidy A,Weckamn G, Shell R, Karwowski W, et al. (2008) Health eزects of nanomaterials: a critical appraisal approach and research to practice. Hum Factors Ergon Manuf 18: 293-341.
15. Boyes WK, Chen R, Chen C, Yokel RA (2012) He neurotoxic potential of engineered nanomaterials. Neurotoxicology 33: 902-910.
16. Maudgalya T, Genaidy A, Weckamn G, Shell R, Karwowski W, et al. (2008) A critical appraisal of epidemiological studies investigating the eزects of uOtrafine particles on human health. Hum Factors Ergon Manuf 18: 358-373.
17. Green CJ, Ndegwa S (2011) Nanotechnology: a review of exposure, health risks and recent regulatory developments. National Collaborating Centre for Environmental Health.
18. Jaques PA, Kim CS (2000) Measurement of total lung deposition of inhaled uOtrafine particles in healthy men and women. Inhal Toxicol 12: 715-731.
19. Hillyer JF (2001) Gastrointestinal persorption and tissue distribution of diزerentOy sized colloidal gold nanoparticles. J Pharm Sci 90: 1927-1936.
20. Rouse JG, Yang Y, Rasmussen JP, Barron AR, Riviere NA (2007) (ز ects of mechanical flexion on the penetration of fullerene amino acid-derivatized peptide nanoparticles through skin. Nano Lett 7: 155-160.
21. Murashov V, Howard J (2007) Biosafety, occupational health and nanotechnology. Appl Biosaf 12: 158-167.
22. National Cancer Institute (2016) Safety of Nanotechnology. National Institutes of Health, US Department of Health and Human Services. USA.
23. Kim IJ (2015) Knowledge Gaps and Research Challenges in the Contemporary Ergonomics. J Ergon 5: 2.
24. Kim IJ (2016) Safety and Health Practices in Food Industry and Ergonomic Interventions. J Ergon 6:1.
25. Kim IJ (2016) Accidents and Accident Prevention in the Agricultural Industry: Ergonomic Engagement. J Ergon 6: 3.
26. Kim IJ (2016) Ergonomic Involvement for Occupational Safety and Health Improvements in the Oil and Gas Industry. J Ergon 6: 3.

27. Kim IJ (2016) Cognitive Ergonomics and Its Role for Industry Safety Enhancements. J Ergon 6: 4.
28. Karwowski W (2006) From past to future: building a collective vision for HFES 2020+. Hum Factor Ergon Soc Bul 49: 1-2.
29. Catherine L'Allain C, Caroly SS, Drais E (2014) Areas of discussion about work, resources for prevention of risks related to nanomaterials. International Conference on Safe production and use of nanomaterials, Nanosafe 2014, Session 11 Risk Management 18-20, Grenoble, France.

5

Ethical and Scientific Issues of Nanotechnology in the Workplace

Paul A. Schulte[1] and Fabio Salamanca-Buentello[2]

[1]National Institute for Occupational Safety and Health, Centers for Disease Control and Prevention, Cincinnati, Ohio, USA
[2]University of Toronto Joint Centre for Bioethics and Canadian Program on Genomics and Global Health, Toronto, Ontario, Canada

ABSTRACT

In the absence of scientific clarity about the potential health effects of occupational exposure to nanoparticles, a need exists for guidance in decision-making about hazards, risks, and controls. An identification of the ethical issues involved may be useful to decision makers, particularly employers, workers, investors, and health authorities. Because the goal of occupational safety and health is the prevention of disease in workers, the situations that have ethical implications that most affect workers have been identified. These situations include the *a*) identification and communication of hazards and risks by scientists, authorities, and employers; *b*) workers' acceptance of risk; *c*) selection and implementation of controls; *d*) establishment of medical screening programs; and *e*) investment in toxicologic and control research. The ethical issues involve the unbiased determination of hazards

Citation: Paul A. Schulte and Fabio Salamanca-Buentello, 2007, Ethical and Scientific Issues of Nanotechnology in the Workplace, Open Access Environ Health Perspect, 2007, 6:5 https://doi.org/10.1289/ehp.9456.

and risks, nonmaleficence (doing no harm), autonomy, justice, privacy, and promoting respect for persons. As the ethical issues are identified and explored, options for decision makers can be developed. Additionally, societal deliberations about workplace risks of nanotechnologies may be enhanced by special emphasis on small businesses and adoption of a global perspective.

Science and technology have identified unique properties in materials with dimensions in the range of 1–100 nm [Health and Safety Executive (HSE) 2004; National Nanotechnology Initiative (NNI) 2004]. These properties may yield many far-reaching societal benefits, but they may also pose hazards and risks. One area of concern about hazards is the workplace—be it a research laboratory, start-up company, production facility, or operation in which engineered nanomaterials are processed, used, disposed, or recycled. These are the workplaces in which some of the first societal exposures to engineered nanoparticles are occurring. Such exposures are likely to be inadvertent and unintended. Despite a conscious effort by governments, corporations, nongovernmental organizations (NGOs), trade associations, academics, and workers to anticipate and address potential workplace hazards [Bartis and Landree 2006; Hett 2004; National Institute for Occupational Safety and Heath (NIOSH) 2006; National Science and Technology Council (NSTC) 2006; Roco and Bainbridge 2003; Scientific Committee on Engineering and Newly Identified Health Risks (SCENIHR) 2005], workers are still likely to be exposed to nanomaterials.

Much research on the ethical aspects of nanotechnology has focused on generalized issues such as equity, privacy, security, environmental impact, and metaphysical applications concerning human–machine interactions (Mnyusiwalla et al. 2003; Moor and Weckert 2004; Singer 2004). No ethics research has been carried out that pertains specifically to the workplace. To help anticipate the impact of nanotechnology, it is important to provide a framework for the ethical and scientific issues involved with nanotechnology in the workplace. Ethical analysis may assure society that the expansive promise of nanotechnology does not conceal hazards and risks for workers. An emerging belief is that nanoscience and technology cannot be based on past practices in which ethical and social reflection is a second step to using newly developed science; rather, ethical reflections must accompany research every step of the way (National Academy of Engineering 2004). Our goal in this paper is to identify ethical issues that are directly related to nanotechnology in the workplace and their implications for workers' health and safety.

FRAMEWORK FOR ETHICAL ASSESSMENT

The framework for considering the ethical issues can be drawn from the work of Gert et al. (1997), Gewirth (1978, 1986), and Schrader-Frechette (1994) as well as from the "principlist" approach of Beauchamp and Childress (1994). The ethical issues that most affect workers in jobs involving nanomaterials are linked to identification and communication of hazards and risks by scientists, authorities, and employers; acceptance of risk by workers; implementation of controls; choice of participation in medical screening; and adequate investment in toxicologic and exposure control research (Table 1). The ethical issues involve the identification and assessment of hazards and risks, nonmaleficence (doing no harm), autonomy (self-determination), justice (fairness in distribution of risks), privacy (in handling of medical information), and respect for persons.

Factual scientific knowledge—which is the basis for ethical decisions about occupational safety and health—may be influenced by biases and values (Kantrowitz 1995). Scientific knowledge is unavoidably value laden. No scientific theory can be considered to be wholly objective, but one theory may be more objective than another (Shrader-Frechette 1994). Underlying the ethical decisions are the way in which nanotechnology is depicted, the potential benefits, and the associated hazards and risks. When information about the hazards of nanoparticles is in doubt, the critical question is where to draw the line about the necessary level of protection and the residual risk at a given level of protection.

Risk assessments are partly subjective and likely to be highly politicized. Thus all risk projections are value laden. No single scenario for describing risks and controls can suffice because of the heterogeneous and developmental nature of nanotechnology. The ethical issues will be specific only for the knowledge base at a given time and for a specified production and use scenario. Researchers have suggested that even with that type of specificity, alternative assessments are needed to capture the ethical and political values that inform policies such as those involving nanotechnology (Schrader-Frechette 2002).

CURRENT STATE OF KNOWLEDGE ABOUT NANOTECHNOLOGY HAZARDS AND RISKS

The way in which nanotechnology is depicted may influence society's reactions to research, development, and prevention and control of potential nanomaterial hazards in the work-place (Berube 2004). The term "nanotechnology" is misleading, since it is not a single technology but a multidisciplinary grouping of physical, chemical, biological, engineering, and electronic processes, materials, applications, and concepts in which size is the defining characteristic (Aitken et al. 2004). However, the issues of size, surface characteristics, durability, chemical composition, and other physiochemical features are not well resolved in the definition. A fuller definition also includes structures with novel properties that can be manipulated on the atomic scale (NNI 2004; Salamanca-Buentello et al. 2005).

Nanoparticles can be considered in at least two broad categories: engineered nanoparticles and incidental (or adventitious) nanoparticles. Engineered nanoparticles are designed with very specific properties. Incidental nanoparticles (natural and anthropogenic) are generated in a relatively uncontrolled manner and are usually physically and chemically heterogeneous compared with engineered nanoparticles (NIOSH 2006). Although the four current major production methods of engineered nanoparticles (gas-phase synthesis, vapor deposition, and colloidal and attrition methods) may expose workers by inhalation, dermal absorption, and ingestion, the amount and likelihood of worker exposure has not been well established. The critical question (based on the little information available) pertains to the assessment of hazards and risks. The unifying theme is that nanoparticles are smaller than their bulk counterparts but have a larger surface area and particle number per unit mass; these characteristics generally increase toxic potential as a result of increased potential for reactivity (Aitken et al. 2004). The application of that theory to the whole of nanotechnology rather than to specific particles and processes may increase rather than decrease the uncertainty about hazards and risks. Increasingly, other characteristics (e.g., surface characteristics) in addition to particle size, that influence toxicity are being identified (Donaldson et al. 2006; Warheit et al. 2004). These characteristics are tremendously variable. Consequently, it is useful to put some limits on the uncertainty by being more precise in the language used to describe nanoparticle hazards and risks. Because a diverse mix of particles and processes exists, hazards and risks are likely to be more accurately assessed on a case-by-case basis—or at least according to the type of production methods and whether particles are embedded in a matrix or unbound.

Knowledge about Hazards and Risks

Health effects data on workers involved with nanotechnology are limited because of the incipient nature of the field, the relatively small number of workers potentially exposed to date, and the lack of time for chronic disease to develop and be detected. The most relevant human experience deals with exposures to ultrafine particles (which include particles with diameters < 100 nm) and fine particles (particles with diameters < 2.5 μm). Ultrafine and fine particles have been assessed in epidemiologic air pollution studies and in studies of occupational cohorts exposed to mineral dusts, fibers, welding fumes, combustion products, and poorly soluble, low-toxicity particulates such as titanium dioxide and carbon black (Maynard and Kuempel 2005; Nel et al. 2006). The hazards of these exposures and exposures to engineered nanoparticles are also identified in animal studies (Donaldson et al. 2004, 2006; Elder et al. 2006; Lam et al. 2004, 2006; Oberdörster et al. 2005; Shvedova et al. 2005; Warheit et al. 2004). A strong relationship exists between the surface area, oxidative stress, and proinflammatory effects of nanoparticles in the lung. The greater the oxidative stress, the more likely the risk of inflammation and cytotoxicity (Nel et al. 2006; Oberdörster et al. 2005). The findings from animal studies ultimately need to be interpreted in terms of the exposure (dose) that humans might receive. Although there is still some debate, the evidence from air pollution studies associates increased particulate air pollution (the finer particulate matter fraction, $PM_{2.5}$, with an aerodynamic diameter < 2.5 μm) with adverse health effects in susceptible members of the population—particularly the elderly with respiratory and cardiovascular diseases [Mark 2004; Peters 2005; U.S. Environmental Protection Agency (U.S. EPA) 2004]. Moreover, the concentrations associated with measurable effects on the health of populations are quite low (Aitken et al. 2004).

In occupational studies, the populations that are repeatedly exposed to hazardous mineral dusts and fibers in the respirable range (e.g., quartz and asbestos, respectively) have well-known health effects related to the dose inhaled (Maynard and Kuempel 2005). With asbestos, the critical risk factors for developing respiratory diseases are fiber length, diameter, and biopersistence. For poorly soluble, low-toxicity dusts such as titanium dioxide, smaller particles in the nanometer size range appear to cause an increase in risk for lung cancer in animals on the basis of particle size and surface area (Heinrich et al. 1995; Oberdörster et al. 2005; Tran et al. 2000).

Although the findings are not conclusive, various studies of engineered nanoparticles in animals raise concerns about the existence and severity of hazards posed to exposed workers (Kipen and Laskin 2005). Possible adverse effects include the development of fibrosis and other pulmonary effects after short-term exposure to carbon nanotubes (Lam et al. 2006; Oberdörster et al. 2005; Shvedova et al. 2005), the translocation of nanoparticles to the brain via the olfactory nerve, the ability of nanoparticles to translocate into the circulation, and the potential for nanoparticles to activate platelets and enhance vascular thrombosis (Radomski et al. 2005).

None of these findings are conclusive about the nature and extent of the hazards, but they may be sufficient to support precautionary action.

Ultimately, the significance of hazard information depends on the extent to which workers are exposed to the hazard. This is the defining criterion of risk (the probability that an exposed worker will become ill). A need has been identified for nanoparticle-specific risk assessments (i.e., those that use the most appropriate dose metrics rather than typical mass) that will be unique to nanotechnology (National Academy of Engineering 2004; SCENIHR 2005).

Risk assessment has been widely used to manage the uncertainty of risks posed to humans by newly introduced chemicals or processes. However, nanotechnology encompasses a diverse range of compositions, structures, and applications, so a single risk assessment and management strategy may not be appropriate (Wardak and Rejeski 2003). Nanotechnology involves the manipulation of matter at the nanoscale to produce materials, structures, and devices that contain various particle types, sizes, surface characteristics, and coatings. These particles may best be addressed by a range of risk assessments specific to the type of particle (composition, surface characteristics, and shape) being assessed. Because of the general inverse relationship between particle size and surface area, dose–effect relationships may vary as a function of total surface area and number of particles rather than mass units (SCENIHR 2005). Risk assessments will be useful to the extent that they reflect the effects of particle sizes and surface area, but such assessments may also need to reflect other particle characteristics. Moreover, it is currently unclear the extent to which the toxico-kinetics (an important component in risk assessment) can be predicted from knowledge of physicochemical properties of nanoparticles (SCENIHR 2005).

Evidence base for Hazard Controls

The most frequently used model of the workplace environment identifies sources of hazards and routes of exposure (e.g., inhalation, skin) [Office of Technology Assessment (OTA) 1985]. Control can be introduced at each of these points. Occupational safety and health professionals have identified a hierarchy of controls based on reliability, efficiency, and the principle that the environment should be controlled before the worker is required to take any preventive action (OTA 1985). In its simplest form, the hierarchy of controls specifies that engineering controls (including substitution, enclosure, isolation, and ventilation) are preferred to the use of personal protective equipment (such as protective clothing and respirators). Work practices are frequently incorporated in risk management efforts to minimize worker exposures, and they often supplement the use of engineering controls. Administrative controls such as worker rotation are sometimes included and generally constitute the "third line of defense" when engineering controls and work practice controls cannot achieve the desired level of worker protection (OTA 1985).

In the absence of adequate toxicity information and extensive history of engineered nanomaterials use, the rationale for control guidance has been based on experience in controlling exposures to incidental ultrafine particles and gases. Airborne nanoparticles are considered to have no inertia—hence, they will behave similarly to gases and will diffuse if they are not fully enclosed (Aitken et al. 2004). A rich history of aerosol science describes the fundamental properties of aerosols and their control [American Conference of Governmental Industrial Hygienists (ACGIH) 2001; Brown 1993; Burton 1997; Davies 1966; Friedlander 1977; Fuchs 1964; Hinds 1999; Ratherman 1996]. Although ultrafine particles are considered equivalent to nanoparticles by some authorities (SCENIHR 2005), they are usually (but not exclusively) at the upper end of the nanoscale range. If airborne nanoparticles conform to the classical physics and aerodynamics observed for larger particles, then controls effective in capturing fine and ultrafine particles and gases (such as source enclosure, local exhaust ventilation, and personal protective equipment) should be effective with the current generation of nanomaterials. It is reasonable to believe that most control methods used for fine and ultrafine particles and also for gases will be useful for controlling nanoparticles, but there is no reason to expect that application of these methods to new nanoparticle generation processes will result in better control than that previously demonstrated for microscale powders and gases (Aitken 2004). A considerable body of opinion indicates that the adverse effects of nanoparticles cannot be predicted (or derived) from the known toxicity of bulk materials with similar chemical composition and surface properties (SCENIHR 2005). Control options for nanoparticles range from no controls to the use of isolation and containments practiced with radiation, gases, and biological agents. The question is where in this continuum should controls be selected. This may also translate into how much money to invest in them. When risks are known to be high or low, the decision is relatively easy, and the appropriate control strategies are generally apparent. However, when hazards are uncertain (as they are with nanoparticles), the difficulty is in deciding what level of controls is warranted (Figure 1). Given the paucity of toxicity information, control guidance must be regarded as interim, and some authorities believe that it should be precautionary—that is, tending toward reducing exposures as much as possible (HSE 2004).

Summary of Evidence on Hazards and Controls

The evidence base pertaining to nanotechnology hazards and controls has been reviewed in various publications (Hett 2004; Maynard and Kuempel 2005; National Academy of Engineering 2004; NIOSH, 2006; Royal Society and Royal Academy of Engineering 2004; SCENIHR 2005) and is summarized in Table 2 by four categories of knowledge described in terms of hazards and controls and awareness. These categories are mutable and pertain to the state of knowledge at a given time. Category 1 ("what we know we know") indicates that we have some knowledge about the health hazards posed by some types of nanoparticles (e.g., ultrafine particles) and gases and how to control them. This category applies to the current generation of engineered nanoparticles and is the basis for much of the current guidance. Category 2 knowledge ("what we know we don't know") is the basis for much of the research currently being conducted or planned. In general, we do not know much about the hazards of new or anticipated engineered particles or whether enough precautions have been taken. A major question is not only how to control exposure but also what are the appropriate extent and cost of controls. Category 3 knowledge ("what we don't know we know") represents the under-utilization of established knowledge. That is, scientists have had extensive experience in hazard and exposure control for ionizing radiation, biological agents, pharmaceuticals, grain and mineral dusts, and air pollution. This experience could be more directly brought to bear on controlling the hazards of nanomaterials in the workplace. In addition, this category could include proprietary information about nanoparticles that is not available for hazard assessments. Category 4 knowledge ("what we don't know we don't know") represents a perennial area of philosophical exploration (Caws 1998). This category includes the range of scenarios about the potency of hazards and the extent of risks. Will new scenarios present new types of exposures and risks? The popular literature on nanotechnology is replete with characterizations of possible future scenarios, but no projections have been made of workplace hazards and risks (Drexler 1986; Regis 1995). Category 4 knowledge also includes the lack of awareness of factors influencing an issue. This lack of awareness can be addressed by engaging a wide variety of disciplines and communities of interest to characterize an issue (HSE 2004). Category 4 knowledge also includes the beliefs we hold that may be wrong. Such beliefs could lead to taking or not taking protective measures on the basis of faulty assumptions. Eventually, Category 4 knowledge can be transformed to Category 2 and then to Category 1.

Regardless of which type of knowledge is considered, the ultimate ethical requirement is to accurately portray the state of knowledge about a hazard or risk and not to understate or overstate it. However, given the developmental nature of nanotechnology, the knowledge of hazard potential will change over time and require restatement and possibly modification of guidance. In the absence of adequate hazard and risk assessment data, the critical question is how much caution is warranted.

ETHICAL ISSUES

Identifying and Communicating Hazards and Risks

The "hazard identification" stage of risk analysis is the basis for risk management decisionmaking. The output of this stage is often highly debated, since the process of reasoning is primarily qualitative and the results trigger other stages of analysis and decisions about preventive action (Crawford-Brown and Brown 1997). Interpreting scientific information about the hazards of nanomaterials is basic to communicating the hazards and risks posed to workers. Interpreting and communicating hazard and risk information is an integral part of risk management by employers. The employers' decisionmaking will focus on deciding which preventive controls should be used to assure a safe and healthful workplace.

Employers and workers look to scientists and authoritative organizations to help interpret hazard and risk information and to put it into context.

This expectation may pressure scientists to go beyond the mere conduct of research. The interface between science and morality is exceedingly complex, but scientists are generally considered to have ethical obligations to society at large (Pimple 2002; Schrader-Frechette 1994; Weil 2002). However, no consensus has been reached about the nature of those ethical obligations beyond fulfilling the professional responsibilities internal to scientific research. Framing a clear and coherent approach to the ethical responsibilities of scientists in nanotechnology is a difficult task. At the least, such an approach requires scientists to use appropriate qualifiers in published papers and to be cautious in generalizing their results. More broadly, it means not shrinking from considering the implications of their work, even if all the scientific details are not known.

Decision makers may have inadequate scientific information to help them decide how precautionary their approach should be (Cairns 2003). To determine whether a decision conforms with the principle of nonmaleficence, decision makers must determine the harm that could occur if the nanoparticles were as toxic as suggested by preliminary hazard information. Data on air pollution and industrial ultrafine particles indicate that a given mass of nanoparticles would be more biologically reactive and hence potentially more toxic than the same mass of larger particles (Seaton 2006). Consequently, the level of control might need to be more stringent for smaller nanoscale dusts than for those with diameters > 100 nm. Ultimately, the more stringent level of controls may result in risks that are equal to or smaller than risks posed by larger particles. Authoritative organizations and employers are responsible for communicating the risk workers face after appropriate controls are implemented. Failure to do so may preclude workers from exercising autonomy. This issue may be confounded by the fact that the employer has a proprietary interest in not releasing information about "nanoproducts" and workplace controls.

The principlist ethical approach focuses on principles such as nonmaleficence and autonomy but fails to assess the social and organizational context of occupational safety and health and the role of practitioners in relation to the corporate structure (Gert et al. 1997; Samuels 2003). With regard to nanotechnology, the contextual pressures on practitioners and authorities arise from a company's or society's needs and desires for nanotechnology to grow and develop. Mention of potential health concerns may be seen as alarmist, unfounded, and detrimental to the growth of the field. Nonetheless, the counter position is that conflicting demands on practitioners from being both an agent of a company and an autonomous professional constitute a social and structural problem rather than a problem of individual ethics (Draper 2003; Samuels 2003). One solution is that health pronouncements be made independently of promotional concerns for nanotechnology.

Workers' Acceptance of Risk

Acceptance of risk is a relative concept that includes judgment about the certainty and severity of risk, the extent of the health effects, voluntary nature of the risk, the risks and advantages of any alternatives, and compensation for undergoing the risk (Fischoff 1994). It is a false premise to assert that workers have free choice in terms of which work and working conditions to accept. Although some component of self-determination is present, economic and social conditions exert the greatest influences on workers' selection of work, level of risk tolerated, and ability to participate in risk management. Worker participation in risk management is not a static concept and has increased over the past 35 years with the implementation of team approaches, management systems, corporate responsibility, and right to know and act movements (Gallagher 1997; Jensen 2002; Lynn 1997; Shearn 2005). Nonetheless, workers generally cannot universally refuse work they consider hazardous and still keep their jobs. Conformance with the principle of autonomy depends on the extent to which workers have input into risk management at their work sites and the degree to which they are at risk after controls have been implemented.

Justice is also related to worker decision-making. At issue is the extent to which workers are exposed to greater risks than the general public—or, stated another way, whether it is appropriate to exchange incentives such as wages or hazardous duty pay for additional risk from exposure to nanoparticles (Schrader-Frechette 2002). This issue may be less significant if nanoparticle controls reduce workers' risk levels to those of the general public, if conceivably both are known. Clearly, society accepts that some jobs are inherently riskier than others. However, in many countries the societal goal is to provide a safe and healthful workplace for all workers.

Selecting and Implementing Controls

The critical ethical question related to control of nanoparticles is whether sufficient controls are being implemented to prevent injury and illness. If not, worker exposures may result in increased risk of harm or actual harm. The central scientific fact is that the risk posed by nanomaterials is not well established. However, preliminary information suggests that at least the same level of concern afforded to industrial fine and ultrafine particles should be extended to engineered nanomaterials and that a commensurate level of protection should be instituted for them (Hett 2004; Royal Society and Royal Academy of Engineering 2004; Seaton 2006). Any risk posed by exposure to ultrafine particles is a function of their potential toxicity and the extent of exposure. Based on limited toxicological evidence of risk and a heightened level of concern, the best approach might be to treat engineered nanoparticles as if they were potential occupational hazards and to use a prudent health-protective, risk-based approach to develop interim precautionary measures consistent with good professional occupational safety and health practice (Royal Society and Royal Academy of Engineering 2004).

Such interim precautionary measures could include guidelines for conducting work-place exposure assessments, implementing engineering controls, designating work practices, and developing process or industry interim exposure limits as core elements. If the focus of exposure control is airborne particles of respirable dimensions, such approaches may be useful and reflect the professional judgment of experienced practitioners. If skin absorption is also a likely route of exposure, guidelines should be developed for preventing skin exposure. Unfortunately, data are insufficient to make a strong risk-based assessment to inform these decisions.

The evidence suggests that at least some manufactured nanoparticles will be more toxic per unit of mass than larger particles of the same chemicals (Royal Society and Royal Academy of Engineering 2004). However, some evidence indicates that with the use of existing controls for fine or ultrafine particles, workers will not be at inordinately elevated risk for lung disease. For example, estimates based on animal studies indicate that workers exposed to ultrafine titanium dioxide at 0.1 mg/m^3 for a 45-year working lifetime have an excess risk of lung cancer that is $< 1/1,000$ and could in fact have a risk approaching zero (Kuempel et al. 2004). The basis for these findings is the hazard posed by increased particle surface area for a given mass of small-sized particles, as derived from animal studies and extrapolated to humans. The extent to which this analysis pertains to other nanoparticles is not known and may vary depending on morphology, surface activity, and biopersistence. Moreover, precise risks from exposure to these ultrafine particles can be determined only if adequate animal or human data are available. Also, if particles can translocate into the central nervous system or the circulatory system, further estimates will be required before conclusions can be drawn (Oberdörster et al. 2005).

In short, given the insufficient evidence of hazards posed by the current generation of nanoparticles, the risks (whatever they may be) are expected to be reduced when controls recommended for known industrial ultrafine particles (such as titanium dioxide) are utilized. This conclusion is supported by *a*) a generalized risk assessment based on surface area for poorly soluble, low-toxicity particles and *b*) the fact that such particles conform to classic physics and aerodynamic laws when airborne. However, future assessments of risk could be different, depending on the bio-persistence, structure, surface activity of new particles, and information about translocation across endothelial cell barriers. If these topics are the focus of risk communications and management efforts, there appears to be general conformance with the ethical principles of beneficence and nonmaleficence. At the same time, no strong evidence indicates that workers in these environments are not at excess risk. Minimal risk is only assumed on the basis of qualitative risk assessments and the utility of proven controls for some types of particles.

Overall, the knowledge base pertaining to nanomaterials is not static but changes as scientists develop new materials and conduct toxicological or other health effects research. Consequently, ongoing evaluation of health risks is needed along with continued communication and development of management plans to be in conformance with the ethical principles discussed in this article.

Establishing Medical Screening Programs

Medical screening is the application of tests to asymptomatic persons to detect those in the early stages of disease or at risk of disease. Medical screening in the workplace differs from medical screening in the general population because of the specific nature of the occupational condition and responsibilities of employers (Halperin et al. 1986; Harber et al. 2003). A wide range of ethical questions has been identified regarding the medical screening of workers and the use and implications of the findings (Ashford et al. 1990; Schulte 1986). These questions address the rationale for screening, the voluntary nature of the screening, the action that will be taken for workers with positive tests, and individuals who will have access to test information.

Medical screening is not generally warranted when the toxicity of a material and the workers' risk are unknown—as is the case with most nanomaterials. Moreover, for diseases such as lung cancer (which is a potential outcome resulting from some nanoparticle exposure), no strong evidence base exists for routine screening; and general population screening for lung cancer is not generally recommended [National Cancer Institute (NCI) 2006]. Not only does screening fail to reduce mortality from lung cancer, it could lead to false-positive tests and unnecessary invasive procedures or treatments (NCI 2006). Medical screening of workers may be warranted for nonmalignant respiratory effects in some nanotechnology operations where significant residual risks may occur after controls are implemented. Such screening should be part of a comprehensive risk management program that considers not only respiratory hazards but also cardiovascular and neurologic risks as well as risks in various other potential target organ systems (Oberdörster et al. 2005; Radomski et al. 2005; Tran et al. 2005). If various nanomaterials are found to have toxic effects and if appropriate (validated) tests exist for early detection of those effects in exposed workers, medical screening might be warranted. However, medical screening is historically viewed as a secondary preventive effort in the hierarchy of controls (Ashford et al. 1990)

The ethical questions that apply to the medical screening of workers pertain to whether the screening is voluntary, who will have access to the results, and what the purpose of such access will be. Screening generally requires diagnostic confirmation; and for positive cases, screening requires timely treatment. Who is financially responsible for these procedures? Ethical issues can also arise in the use of screening results to label or stigmatize workers or to remove them from a job. Screening results may also create psychological burdens. Resolving such ethical issues will depend partly on the degree to which the worker has been informed about how the results will be used.

Ensuring Adequate Investment in Toxicological and Control Research

Ethical issues cannot be adequately addressed for nanotechnology without sufficient knowledge of the hazards involved. Because limited information is available on the safety of an ever-growing number of nanomaterials, an ongoing research effort is needed to comport with the principles of autonomy, beneficence, and nonmaleficence. In addition, research is needed on the extent of exposure and the effectiveness of controls. Internationally, such research is under way.

However, the question of the level of funding of this research has ethical implications because much of the current control guidance is precautionary and is not based on strong quantitative risk assessments. Further research is the only way to address this lack of appropriate information.

Some commentators have called for a slowdown in research and development of nanoparticles, whereas others have identified a need for increased health effects research and ethical analysis [Action Group on Erosion, Technology and Concentration (ETC Group) 2003, 2004; Mnyusiwalla et al. 2003]. The needs for health-based research have been identified and include the following topics: exposure and dose, toxicity, metrology, epidemiology, control technology, safety, education, recommendations, and applications in the near term (NIOSH 2006).

Researchers could help further the discussions of ethical issues by assessing the global budget for nanotechnology research and development and by determining the actual amounts dedicated to occupational safety and health research and ethical research in this field. Globally, such information is not well documented; but existing U.S. data can be considered. For the first time since the inception of the NNI, funding for 2005 was classified by program component area. The funding for the Societal Dimensions component area included $US39 million for environment, health, and safety and $43 million for educating the public about the broad implications of nanotechnology for society (including economic, workplace, education, ethical, and legal implications). This funding came from 11 agencies with a combined nanotechnology budget of approximately $1.054 billion. The level of funding (7.8% of the total) has been criticized as insufficient for the societal dimensions component and the subset dedicated to occupational safety and health (Bartis and Landree 2006; Maynard 2006; Service 2005). Nonetheless, there is a concerted international effort to address health and safety aspects of nanomaterials (NSTC 2006; Thomas et al. 2006).

Promoting Respect for Persons

Underlying the debates about nanotechnology has been the issue of tolerating the potential for harm to some in the context of anticipated benefits to society. Such thinking embodies the utilitarian point of view that harm to one person may be justified by a larger benefit to someone else (Harris 2003). This point of view contrasts with the ethical principle of respect for persons, which emphasizes the rights of the individual and is associated with the golden rule ("Do unto others as you would have them do unto you") (Gewirth 1978, 1986). In the workplace, this principle translates to acknowledging for each worker the right to a safe and healthful work environment. This right imposes correlative duties on the employers and governments who must secure the workers' rights to a safe and healthful workplace (Gewirth 1986). The objection to this interpretation is that the rights of employers, and hence the rights of society, to property and benefit resulting from nanotechnology may be (or may appear to be) in conflict with workers' rights. When two rights conflict with each other, some rational way must be found to determine their relative priority. Gewirth (1986) identified an essential criterion for such priority as the degrees of necessity for action. For example, where the property rights of employers may be in conflict with workers' rights to safety and health, the diminution of health or a threat to safety lowers one's capacity for action and is a greater loss than some decrease in another's property, wealth, or freedom to control it. The practical implication is this: In the absence of adequate information about nanotechnology hazards, risks, and controls, employers should be moved to use more rather than fewer control measures (Hett 2004). Conducting site-specific hazard assessments and using appropriate controls appear to demonstrate conformance with the principle of respect for persons and with the principles of autonomy, beneficence, and nonmaleficence. However, the extent of control measures required may be the key matter of dispute. For the most part, control of the current generation of most engineered nanoparticles is within the capabilities of existing technologies. The issue is how much to invest in applying those technologies in a given workplace.

STRATEGIES FOR SUPPORTING ETHICAL DECISIONMAKING

Placing Special Emphasis on Small Businesses

The occupational safety and health problems of small businesses have been a major focus of concern, particularly in the last decade, since most workplaces are classified as small (i.e., workplaces that employ fewer than 250, 100, or 20 workers, depending on the definition). This statement is likely to hold true for work-places involving nanotechnology, but it is not well documented (Aitken et al. 2004; Roco and Bainbridge 2003).

The frequency of occupational injury and illness in small businesses may exceed the average for general industry across all businesses in a sector, but the frequency may not be evident in an individual company (NIOSH 1999). Small businesses are generally perceived to have little time and few resources dedicated to occupational safety and health.

Small businesses are the driving force of most economies, including the subset of economies related to nanotechnology (Roco and Bainbridge 2003). Independent consultants, trade associations, insurance companies, product suppliers, and government agencies are the major sources of occupational safety and health information for small businesses. Occupational safety and health information may also be passed to downstream users of nanoparticles from upstream suppliers. In fact, for documented hazards, suppliers may have an ethical or legal obligation to pass on such information to downstream customers. There is a need for occupational safety and health guidance information about nanotechnology hazards and controls for small businesses.

Adopting a Global Perspective

The growth of nanotechnology is a global phenomenon that requires a global approach to hazards and risks, particularly in the workplace. The world needs internationally valid standards for nanotechnology materials as well as a uniform nomenclature (American Society for Testing and Materials 2005; Hett 2004). Without a uniform nomenclature, investigators, insurers, regulators, governments, companies, and workers could have difficulty communicating and taking concerted actions.

The flow of materials in the global economy crosses many borders, including those of developing nations (Salamanca-Buentello et al. 2005). Thus, to assure the safety and health of workers, decision makers (whether they are employers or government authorities) must know and understand what materials are used in various processes and operations. This issue is complicated because many different definitions and descriptions may be used in science-based and regulation-based documents. To develop nanotechnology with minimal risks, knowledge gaps must be identified and addressed through international cooperation. Also needed is a transparent risk assessment framework that can achieve wide acceptability (SCENIHR 2005).

Global approaches to sharing occupational safety and health information require increased opportunity and capacity to access information. The "right" to know about risks—or more broadly, the right to information—is not evenly recognized worldwide (Pantry 2002). The World Health Organization (WHO) promotes the right to health at work for all. Information is a means to realizing that right. Despite broad WHO membership by many countries, true access to information and distribution within countries is still a problem.

Risk communications (including material safety data sheets) should reflect a degree of uniformity worldwide. International collaboration is warranted to ensure that hazardous processes are not relegated to countries with cheap labor markets or lax environmental controls (Singer et al. 2005, 2006). A critical issue that has both national and global implications is whether countries will treat nanomaterials made of a given substance differently from materials made with larger particles of the same substance. The characteristics of nanoparticles may be different from those of the larger particles with the same composition. For example, most materials made from carbon generally appear to pose a minimal health risk; however, nanotubes made of carbon may pose a greater health risk yet be regulated at the less protective level (Shvedova et al. 2005). The issue is whether to recommend the same risk communication and management strategy for both. On the basis of the carbon nanotube example, new standards and risk communication materials are likely to be required for at least some nanoparticles.

CONCLUSIONS

The ethical questions about nanotechnology in the workplace arise from the state of knowledge about the hazards of nanomaterials and the risks they may pose to workers. The lack of clarity on these issues requires an interim assessment of the hazards and risks that might exist in various situations. Workers will be able to exercise their autonomy only if the processes leading to hazard identification and risk assessment are transparent and understandable. Employers will conform to the principles of autonomy, beneficence, nonmaleficence, justice, privacy and respect for persons to the extent that they *a*) accurately portray hazards and risks, *b*) are precautionary in their approach to hazards, *c*) engage in communication and dialogue with workers, and *d*) take the necessary steps to control risks so that they appear reasonable and acceptable to workers.

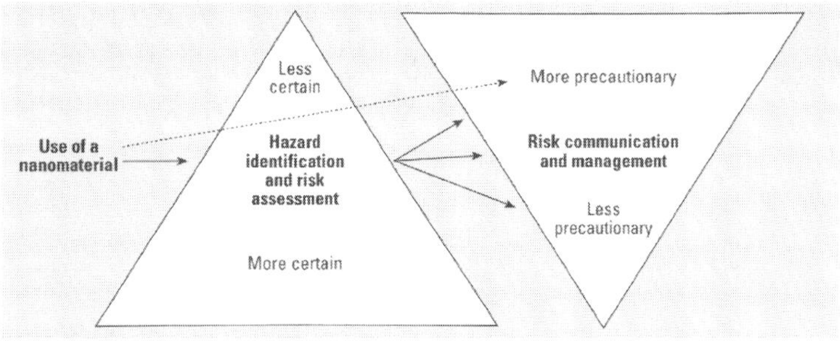

Figure 1: Risk management decisionmaking for nanoparticles in the workplace: what is the appropriate level of controls?

Table 1: Ethical issues pertaining to workplace situations involving nanomaterials

Work-related scenarios	Ethical principles involved	Decision making issues
Identification and communication of hazards and risks	Responsibilities of scientists Nonmaleficence Autonomy Respect for persons	Extent to which strengths and weaknesses of data are identified Degree of participation in public discussion Accuracy of communications Timeliness of communications
Workers' acceptance of risks	Autonomy Respect for persons Justice	Extent of inclusion of workers in decision making
Selection and implementation of workplace controls	Nonmaleficence Beneficen-ceRespect for persons	Level of control technologies utilized
Medical screening of nanotechnology workers	Autonomy Privacy Respect for persons	Appropriateness of the rationale for medical screening Extent to which participation is voluntary Maintenance of privacy test results
Investment in toxicological and control research	Nonmaleficence Justice Respect for persons	Adequacy of investment

Table 2: Summary of the state of knowledge for nanoparticle hazards and controls

Awareness of knowledge	Content of knowledge (hazards and controls)
1. What we know we know	Health effects of ultrafines, air pollution, and fibers How to control ultrafine particles in the workplace Importance of size, surface area, and surface characteristics Serious health effects of some nanoparticles in animals Translocation of some nanomaterials along the olfactory nerve in animals
2. What we know we don't know	Measurement and characterization techniques Hazards of newly engineered particles Extent of translocation in the body Interaction with contaminants in the workplace Importance of dermal exposure Health effects in workers Risks to workers Effectiveness of controls Advisability of medical screening and biological monitoring Risk to workers' families
3. What we don't know we know	Extensive experience available in controlling hazardous substances and agents (radiation, biological agents, pharmaceuticals) that can be applicable to nanoparticles Proprietary nanoparticle information Lessons from previous "new" technologies
4. What we don't know we don't know	Unanticipated new hazards Unanticipated new controls Wrong assumptions about hazards and controls

We thank the following for input or comments on earlier drafts: M. Ellenbecker, S. Samuels, H. Kipen, M. Hoover, E. Kuempel, R. Zumwalde, C. Geraci, V. Murashov, P. Middendorf.

The findings and conclusions expressed in this paper are those of the authors and do not necessarily represent the views of the National Institute for Occupational Safety and Health.

REFERENCES

1. ACGIH. 2001. Industrial Ventilation: a Manual of Recommended Practice. Cincinnati, OH:American Conference of Governmental Industrial Hygienists.
2. Aitken RJ, Creely KS, Tran CL. 2004. Nanoparticles: An Occupational Hygiene Review. Health Safety Executive, Research Report 274. London:HSE Books.
3. Ashford NA, Spadafor CJ, Hattis DB, Caldart CC. 1990. Monitoring the Worker for Exposure and Disease: Scientific Legal and Ethical Considerations in the Use of Biomarkers.
4. Baltimore, MD:The Johns Hopkins University Press. American Society for Testing and Materials. 2005. New Standard Terminology for Nanotechnology. WK8051. Available: http:// www.astm.org/cgi-bin/SoftCart.exe/DATABASE.CART/ WORKITEMS/WK8051.htm? L+mystore+rdik5786+1158038968 [accessed 11 September 2006].
5. Bartis JT, Landree E. 2006. Nanomaterials in the Workplace: Policy and Planning Workshop on Occupational Safety and Health. Arlington, VA:RAND Corporation.
6. Beauchamp TL, Childress JI. 1994. Principles of Biomedical Ethics. 4th ed. New York:Oxford University Press.
7. Berube DM. 2004. The rhetoric of nanotechnology. In: Discovering the Nanoscale (Baird D, Nordman A, Schummer J, eds). Amsterdam:IOS Press, 173–192.
8. Brown RC. 1993. Air Filtration. New York:Pergamon Press.
9. Burton J. 1997. General methods for the control of airborne hazards. In: The Occupational Environment—Its Evaluation and Control (DiNardi SR, ed). Fairfax, VA: American Industrial Hygiene Association.
10. Cairns J Jr. 2003. Integrating top-down/bottom-up sustainability strategies: an ethical challenge. Ethics Sci Environ Politics 1–6. Available: http://www.int-res.com/articles/esep/2003/ E26. pdf#search=%22cairns%20integrating%20topdown%202003%22 [accessed 11 September 2006].
11. Caws P. 1998. Communication lag. Sci Commun 20:14–20.
12. Crawford-Brown DJ, Brown KG. 1997. A framework for assessing the rationality of judgments in carcinogenicity hazard identification. Risk 8.307. Available: http://www.piercelaw.edu/ risk/vol8/fall/Cr-Br+.htm [accessed 11 September 2006].

13. Davies CN. 1966. Aerosol Science. London:Academic Press.
14. Donaldson K, Aitken R, Tran L, Stone V, Duffin R, Forrest G, et al. 2006. Carbon nanotubes: a review of their properties in relation to pulmonary toxicology and workplace safety. Toxicol Sci 92:5–22.
15. Donaldson K, Stone V, Tran CL, Kreyling W, Borm PJA. Nanotoxicology. 2004. Occup Environ Med 61:727–278.
16. Draper E. 2003. The Company Doctor: Risk and Responsibility and Corporate Professionalism. New York:Russell Sage Foundation.
17. Drew S. 1999. Building knowledge management into strategy: making sense of a new perspective. Long Range Planning 32:130–136.
18. Drexler KE. 1986. Engines of Creation. New York:Anchor Press/ Doubleday.
19. Elder A, Gelein R, Silva V, Feikert T, Opanashuk L, Carter J, et al. 2006. Translocation of inhaled ultrafine manganese oxide particles to the central nervous system. Environ Health Perspect 114:1172–1178.
20. ETC Group. 2003. The Big Down: From Genomes to Atoms. Winnipeg, Ontario, CN:Action Group on Erosion, Technology and Concentration.
21. ETC Group (Action Group on Erosion, Technology and Concentration). 2004. Down on the Farm: The Impact of Nano-Scale Technologies on Food and Agriculture. http://www. etcgroup.org/upload/ publication/80/01/etc_dotfarm2004.pdf [accessed 16 November 2006].
22. Fischoff B. 1994. Acceptable risk: a conceptual proposal. Risk 5.1. Available: http://www.piercelaw.edu/risk/vol5/winter/ Fischhof.htm [accessed 11 September 2006].
23. Friedlander SH. 1997. Smoke, Dust, and Haze: Fundamentals of Aerosol Behavior. New York:John Wiley and Sons. Fuchs NA. 1964. The Mechanics of Aerosols. Oxford, UK:Pergamon Press.
24. Gallagher C. 1997. Types of health and safety management systems. In: Health and Safety Management Systems: An Analysis of System Types and Effectiveness. Canberra, ACT:Office of the Australian Safety and Compensation Council.
25. Gert B, Culver CM, Clouser, KD. 1997. Bioethics: A Return to Fundamentals. New York:Oxford University Press.
26. Gewirth A. 1978. Reason and Morality. Chicago:University of Chicago Press.

27. Gewirth A. 1986. Human rights in the workplace. Am J Ind Med 9:31–40.

28. Halperin WE, Ratcliffe J, Frazier TM, Wilson L, Becker SP, Schulte PA. 1986. Medical screening in the workplace: proposed principles. J Occup Med 28:547–552.

29. Harber P, Conlon C, McCunney RJ. 2003. Occupational medical surveillance. In: A Practical Approach to Occupational and Environmental Medicine (McCunney RJ, ed). Philadelphia: Lippincott Williams and Wilkins, 582–599.

30. Harris CE. 2003. Methodologies for case studies in engineering ethics. In: Emerging Technologies and Ethical Issues in Engineering. Washington, DC:National Academy of Sciences, 79–93.

31. Heinrich U, Fuhst R, Rittinghauseen S, Creutzenberg O, Bellmann B, Koch W, et al. 1995. Chronic inhalation exposure of Wistar rats and 2 different strains of mice to dieselengine exhaust, carbon black, and titanium dioxide. Inhal Toxicol 7:533–556.

32. Hett A. 2004. Nanotechnology: Small Matter, Many Unknowns. Zurich:Swiss Reinsurance Company. Available: http://www.swissre.com/INTERNET/pwsfilpr.nsf/vwFilebyIDKEYLu/ULR68AKZF/$FILE/Publ04_nano_en.pdf [accessed 16 November 2006].

33. Hinds WC. 1999. Aerosol Technology. New York:John Wiley & Sons.

34. HSE. 2004. Nanotechnology HSE Information Note. Horizons Scanning Information Note no. HSIN1. London:Health and Safety Executive.

35. Jensen PL. 2002. Assessing assessment: the Danish experience of worker participation in risk assessment. Econ Ind Democracy 23:201–207.

36. Kantrowitz A. 1995. The separation of facts and values. Risk 6:105. Available: http:///www.piercelaw.edu/Risk/vol6/spring/ Kantro2.htm [accessed 12 September 2006].

37. Kipen HM, Laskin DL. 2005. Smaller is not always better: nanotechnology yields nanotoxicology. Am J Physiol Lung Cell Mol Physiol 289(5):L696–L697.

38. Kuempel ED, Wheeler M, Smith R, Bailer J. 2004. A quantitative risk assessment in workers using rodent dose-response data of fine and ultrafine titanium dioxide. In: Report of Presentations of Plenary and

Workshop Sessions and Summary of Conclusions. Nanomaterials—A Risk to Health at Work? 12-14 October 2004, Buxton, UK. Buxton, UK: Health and Safety Laboratory. Available: http://www.hsl. gov.uk/capabilities/nanosymrep_final.pdf [accessed 12 September 2006].

39. Lam CW, James JT, McCluskey R, Hunter RL. 2004. Pulmonary toxicity of single-wall carbon nanotubes in mice 7 and 90 days after intratracheal instillation. Toxicol Sci 77:126–134.
40. Lam CW, James JT, McCluskey RL, Arlli S, Hunter RL. 2006. A review of carbon nanotube toxicity and assessment of potential occupational and environmental health risks. Crit Rev Toxicol 36:159–217.
41. Lynn FM. 1997. Public Participation in Risk Management Decisions: The Right to Define, the Right to Know, and the Right to Act. Risk 1.95. Available: http://www.piercelaw.edu/ risk/vol1/spring/lynn.htm [accessed 11 September 2006].
42. Mark D. 2004. Nanomaterials—A Risk to Health at Work? Buxton, UK:Health and Safety Laboratory, SK17 9 JN. Available: www.hsl. gov.uk/capabilities/nanosymrep_ final.pdf [accessed 11 September 2006].
43. Maynard AD. 2006. Nanotechnology: A Research Strategy for Addressing Risk. Washington, DC:Woodrow Wilson International Center for Scholars.
44. Maynard AD, Kuempel ED. 2005. Airborne nanostructured particles and occupational health. J Nanoparticles Res 7:587–614.
45. Mnyusiwalla A, Daar AS, Singer PA. 2003. "Mind the Gap": science and ethics in nanotechnology. Nanotechnology 14:R9–R13.
46. Moor J, Wechert J. 2004. Nanoethics: assessing the nanoscale from an ethical point of view. In: Discovering the Nanoscale (Baird D, Nordmann A, Schummer J, eds). Amsterdam:IOS Press, 301–310.
47. National Academy of Engineering. 2004. Emerging Technologies and Ethical Issues in Engineering: Papers from a Workshop, 14–15 October 2003, Washington, DC. Washington, DC:National Academies Press. Available: http://newton.nap. edu/books/030909271X/html [accessed 16 November 2006]. NCI (National Cancer Institute). 2006. Lung Cancer (PDQ®): Screening. Available: http://www.cancer.gov/cancertopics/ pdq/screening/lung/healthprofessional [accessed 16 November 2006].

48. Nel A, Xia T, Mädler L, Li N. 2006. Toxic potential of materials at the nanolevel. Science 311:622–627.
49. NIOSH. 1999. Identifying High-Risk Small Business Industries: The Basis for Preventing Occupational Injury, Illness, and Fatality. NIOSH (DHHS) Publ no. 99–107. Cincinnati, OH:National Institute for Occupational Safety and Health.
50. NIOSH. 2006. Approaches to Safe Nanotechnology: An Information Exchange with NIOSH. Cincinnati, OH: National Institute for Occupational Safety and Health, Available: http://www.cdc.gov/niosh/topics/nanotech/safenano/ [accessed 11 September 2006].
51. NNI (National Nanotechnology Initiative). 2004. NNI Strategic Plan. Available: www.nano.gov/html/about/strategicplan2004.html [accessed 12 September 2006].
52. NSTC. 2006. The National Nanotechnology Initiative. Environmental, Health and Safety Research for Engineered Nanoscale Materials. Arlington, VA:National Nanotechnology Coordination Office, National Science and Technology Council.
53. Oberdörster G, Oberdörster E, Oberdörster J. 2005. Nanotoxicology: an emerging discipline evolving from studies of ultrafine particles. Environ Health Perspect 113:823–837.
54. OTA. April 1985. Preventing Illness and Injury in the Workplace. OTA-H-256. Washington, DC:U.S. Congress, Office of Technology Assessment.
55. Pantry S. 2002. The Unequal OSH (Occupational Safety and Health) World of Information. Available: http://www. sheilapantry.com/oshworld/editorial/2002/200206.html [accessed 16 November 2006].
56. Peters A. 2005. Particulate matter and heart disease: evidence from epidemiological studies. Toxicol Appl Pharmacol 207:S477–S482.

6

Management of Nanomaterials Safety in Research Environment

Amela Groso[1] , Alke Petri-Fink[2] , Arnaud Magrez[3] , Michael Riediker[4] , and Thierry Meyer[1]

[1] Occupational Safety and Health, School of Basic Sciences, Ecole Polytechnique Fédérale de Lausanne Switzerland
[2] Advanced Particles Group, Department of Chemistry, University of Fribourg, Switzerland
[3] Laboratory of Nanostructures and Novel Electronic Materials, Ecole Polytechnique Fédérale de Lausanne, Switzerland
[4] Institute for Work and Health (Institut universitaire romand de Santé au Travail), Lausanne, Switzerland

ABSTRACT

Despite numerous discussions, workshops, reviews and reports about responsible development of nanotechnology, information describing health and environmental risk of engineered nanoparticles or nanomaterials is severely lacking and thus insufficient for completing rigorous risk assessment

Citation: Amela Groso, Alke Petri-Fink, Arnaud Magrez, Michael Riediker and Thierry Meyer, Management of nanomaterials safety in research environment, Open Access Particle and Fibre Toxicology, 2010, 7:40, https://doi.org/10.1186/1743-8977-7-40.

on their use. However, since preliminary scientific evaluations indicate that there are reasonable suspicions that activities involving nanomaterials might have damaging effects on human health; the precautionary principle must be applied. Public and private institutions as well as industries have the duty to adopt preventive and protective measures proportionate to the risk intensity and the desired level of protection. In this work, we present a practical, 'user-friendly' procedure for a university-wide safety and health management of nanomaterials, developed as a multi-stakeholder effort (government, accident insurance, researchers and experts for occupational safety and health). The process starts using a schematic decision tree that allows classifying the nano laboratory into three ***hazard*** classes similar to a control banding approach (from *Nano 3* - highest hazard to *Nano1* - lowest hazard). Classifying laboratories into ***risk*** classes would require considering actual or potential exposure to the nanomaterial as well as statistical data on health effects of exposure. Due to the fact that these data (as well as exposure limits for each individual material) are not available, risk classes could not be determined. For each hazard level we then provide a list of required risk mitigation measures (technical, organizational and personal). The target ‹users› of this safety and health methodology are researchers and safety officers. They can rapidly access the precautionary hazard class of their activities and the corresponding adequate safety and health measures. We succeed in convincing scientist dealing with nano-activities that adequate safety measures and management are promoting innovation and discoveries by ensuring them a safe environment even in the case of very novel products. The proposed measures are not considered as constraints but as a support to their research. This methodology is being implemented at the Ecole Polytechnique de Lausanne in over 100 research labs dealing with nanomaterials. It is our opinion that it would be useful to other research and academia institutions as well.

Keywords: Precautionary Principle, Personal Protective Equipment, Hazard Level, Asbestos Fiber, Hazard Class

INTRODUCTION

In the last years, nanotechnology has become a key word of public interest, since it brings together different areas of science and benefits from an interdisciplinary or "converging" approach and is expected to lead to innovations that can contribute to addressing many of the problems facing today's society. A scientific and technical revolution has begun that is based upon the ability to systematically organize and manipulate matter on the nanometer length scale. Several nanotechnology-based products have been marketed including electronic components, scratch-free paint, sports equipment, wrinkle- and stain-resistant fabrics, sun creams, and medical products (e.g. bandages, heart valves, MRI contrast agents). Analysts estimate that the market for such products is currently around hundred of billions of euro and could rise to one trillion by 2015 [1]. Accordingly, potential occupational and public exposure to manufactured nanoparticles will increase dramatically in the near future. Many researchers have addressed the toxicity issues associated with different nanoparticles in vitro and in vivo [2, 3, 4]. However, information describing the relative health and environmental risk assessment of engineered nanoparticles or nanomaterials (hereafter, we will use ISO/TR 12885 definition [5] of engineered nanomaterials) is severely lacking. Effects of nanoparticle properties on the immune system are still being explored, and studies of many nanoparticle preparations generally fall into two categories: (a) responses to nanoparticles that are specifically modified to stimulate the immune system or (b) undesirable side-effects of nanoparticles [6]. One initiative that tried to shed light on this issue is a recently completed global review of completed and nearly completed environment, health and safety research on nanomaterials and nanotechnology [7]. The resulting EMERGNANO report is a unique attempt to identify and assess worldwide progress in relation to nanotechnology risk issues. There is no doubt a consensus among producers and users that there is "a need for better characterization of nanotechnology constructs" and for the production of "reagent-grade" nanomaterials, which permit comparison between researches and tests [8]. Recently, this lack of progress in nanotoxicology came under the spotlight [9] again when researchers reported that nanoparticles had been found in the lungs of seven women who had become ill after working in a paint factory in China; two of them later died [10]. However, it remains unclear if the illnesses were caused by the nanoparticles or other chemicals [11]. There is also a widespread agreement [12] that this tragic accident could have been prevented by proper health and safety procedures - the women only occasionally wore masks and the first symptoms appeared five months after the ventilation unit in the factory broke down. At the very least what happened in China emphasises the importance of proper risk management when workers are exposed to nanoparticles for prolonged periods [9].

Initial safety and health strategies [13, 14, 15] were analogies to those for chemicals and powders. Yet, they are not applied consistently and users and producers seem to rely mostly on personal protective equipment [16]. More recent efforts aim at developing strategies that target nano-specific aspects [17, 18, 19, 20]. Recently, the Swiss Government published a precautionary matrix that allows an initial assessment of the risks of nanomaterial applications without requiring detailed knowledge on the toxicology of the nanomaterials involved [21]. Such preliminary information is essential for simplified, so-called control-banding approaches that group risks in broad classes and then define different levels (or bands) of protection efforts [22].

A professional **hazard** is any potential source of damage, injury or harmful effect in respect of a thing or person in certain conditions of the workplace. A hazard is a characteristics of something (tool, machine, product, but also instruction, activity, organization etc..) that can negatively affect the integrity of a person or thing.

Risks are associated with the nature of material and exposures that people have to that material. For a full risk assessment, detailed information is required about the material (chemical composition of nanoparticles, Material Safety Data Sheet when available, particle morphology, aspect ratio, particle size distribution, zeta potential, solubility, known hazards) as well as about the full process descriptions where nanomaterials are used or produced. For each uptake route (respiratory organs, skin, gut) type and level of exposure need to be investigated for each process step. This is a challenge, because it is unclear which characteristics drive the toxicity of nanomaterials and thus need to be measured. Several methods already exist to measure nanoparticle concentrations in air. Mobility particle sizers usually provide reliable and comparable data [23]. However, they are large, expensive and require extensive training, inducing hindrances for routine exposure assessment of workers or researchers.

All this complicates risk assessment considerably. Nevertheless, given the lack of current knowledge about the toxicity of nanomaterials, the difficulty to compare the results obtained from various investigations, and the concern that the bulk materials' safety data sheets might not adequately reflect the real hazardousness of nanomaterials, precaution recommends that all nanomaterials shall be considered potentially hazardous unless sufficient information to the contrary is obtained.

Precautionary Principle

At the Rio Conference on the Environment and development in 1992, world leaders agreed on the Precautionary principle stated in the following terms: 'In order to protect the environment, the precautionary principle shall be widely applied by states according to their capability. Where there are threats of serious or irreversible damage, lack of full scientific certainty shall not be a reason for postponing cost-effective measures to prevent environmental degradation' [24]. While this principle has primarily been used internationally around environmental health issues, other groups are adopting this philosophy to protect the health of workers. In 1996, the American Public Health Association passed a resolution entitled, "The Precautionary Principle and Chemical Exposure Standards for the Workplace". This resolution recognized the need for implementing the precautionary approach, where chemicals are considered potentially dangerous, until the extent of its toxicity is sufficiently known, and the establishment of strict, preventive chemical exposure limits. In February 2000, the European Commission published a Commission Communication on the precautionary principle (EU Resolution on the Precautionary Principle, 2000) providing a general framework for its use in EU policy [25].

So, if a preliminary scientific evaluation emphasizes that there are reasonable grounds for concern that a particular activity might lead to damaging effects on the environment, or on human, animal or plant health, the precautionary principle is triggered. Within this context, we consider the precautionary principle as directly applicable to emerging nanotechnologies.

Being responsible for safety and health in an research institution, we had to determine which actions should be taken, potential effects of taking no actions, the uncertainties inherent in the scientific evaluation, and the views on how to manage the risks. The adopted measures had to be proportionate to the level of risk and the desired level of protection and will evolve with forthcoming knowledge. It is possible that level of protection may be eased, especially for Nano 3 as defined here, as more is known about specific nanomaterials.

Objectives and Motivation

At EPFL (Ecole Polytechnique Fédérale de Lausanne), over 30 research groups (in basic sciences, engineering or life sciences) produce, modify or use engineered nanomaterials (Figure 1) in approximately 100 laboratories with over 300 different associated production or characterization processes. Classical risk assessment methods (Hazard and Operability Studies (HAZOP) - often used to analyze risks in chemical processes, Failure Mode and Effects Analysis (FMEA) - often used in industry to evaluate the effects of potential failure modes, etc.) would require around 2000 man/day workload - huge resources, nearly impossible to obtain. Some other institutions have already developed best practices guides and safety management procedures for nanomaterials [26, 27, 28, 29]. However, they mainly propose a risk analysis approach for each individual process and particle type, which is not very practical for large research centers with many different, constantly changing forms of nano-related studies and laboratories. Or alternatively, good laboratory practices are proposed [30, 31] that apparently are not well respected. Published in the February issue of Nature Nanotechnology [32], Jesus Santamaria and his team have conducted an online survey to identify what safety practices researchers are following in their laboratories. The responses of the 240 participants shed some light on what is going on. The questions covered: details of the materials and processing methods used; safety measures; waste disposal procedure, and knowledge of legislation for handling nanomaterials. One of the most surprising results [33] is that nearly three quarters of respondents reported not having internal rules to follow regarding the handling nanomaterials (approximately half of them didn't have rules and over a quarter were not aware of any internal regulations). All this led to the development of a methodology and procedure helping to answer questions related to safety and health for present and future users of nanomaterials in university setting. The methodology was introduced and tested for applicability at the EPFL.

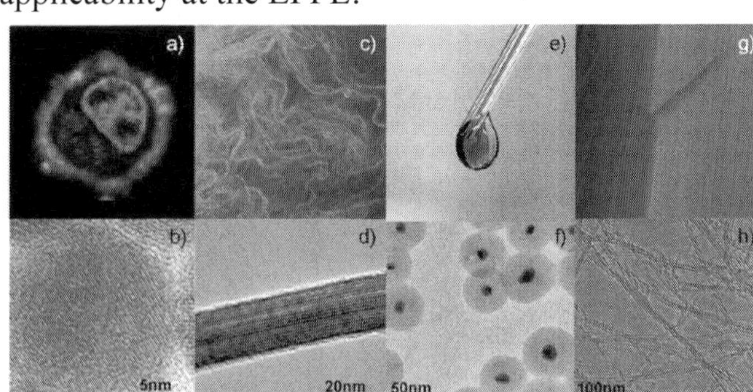

Figure 1: Different forms of nanomaterials produced at the EPFL. Examples of different forms of nanomaterials produced at the EPFL. a) Confocal micrograph of cells exposed to Cy3.5 labelled nanoparticles (red). b) High

resolution transmission electron micrograph of superparamagnetic iron oxide nanoparticles. c) Optical micrograph of V_2O_5 nanowires [47]. d) Transmission electron micrograph of V_2O_5 nanowires. e) Suspension of gold nanoparticles. f) Transmission electron micrograph of gold/silica core-shell nanoparticles. g) Scanning Electron Micrograph of aligned carbon nanotubes forest [48] h) Transmission Electron Micrograph of carbon nanotubes produced by Chemical Vapour Deposition.

Development of Safety Procedure for Nanoobjects Production/ Use

A Nanosafe team consisting of three safety and health specialists, one nano-health and occupational hygiene expert, one insurance representative, three EPFL scientists and nanoparticles' users (production and use) and one representative of State Secretariat for Economic Affairs was appointed to develop a procedure for managing the occupational safety and health risks relevant to research laboratories producing and using nanomaterials. The procedure consists of two parts. Using a decision tree we sort the "nano-laboratories" into three hazard classes (Nano 3 = highest hazard to Nano 1 = lowest hazard), which corresponds to analogue approaches applied to other hazard types (biohazard, radioprotection or chemistry). We then provide a list of required prevention/protection measures (safety barriers) for each hazard level. The target users of this safety and health methodology are at first researchers. They can rapidly access the hazard class of their activity and the corresponding adequate safety and health measures. More detailed analysis of specific activities can be undertaken by safety and health experts when needed. According to our opinion and experience, the proposed management of nanomaterial safety is not stifling or harming innovation, as it is sometimes feared among researchers [34].

Decision Tree for Laboratory Type Determination

Figure 2 depicts the questions to be answered by nanomaterial users and producers (only research environment is considered, industrial processes are not discussed) when classifying their activities. Exposure to nanomaterials may happen by ingestion, inhalation, injection and dermal contact. The main occupational exposure routes are the respiration tract and the skin. Consequently, the first differentiation regards the environment, whether the process is carried out in a closed (complete process confinement) or open system. In case the process is not fully enclosed (glove box or completely sealed environment), different types of activities with nanomaterials are subsequently discussed individually.

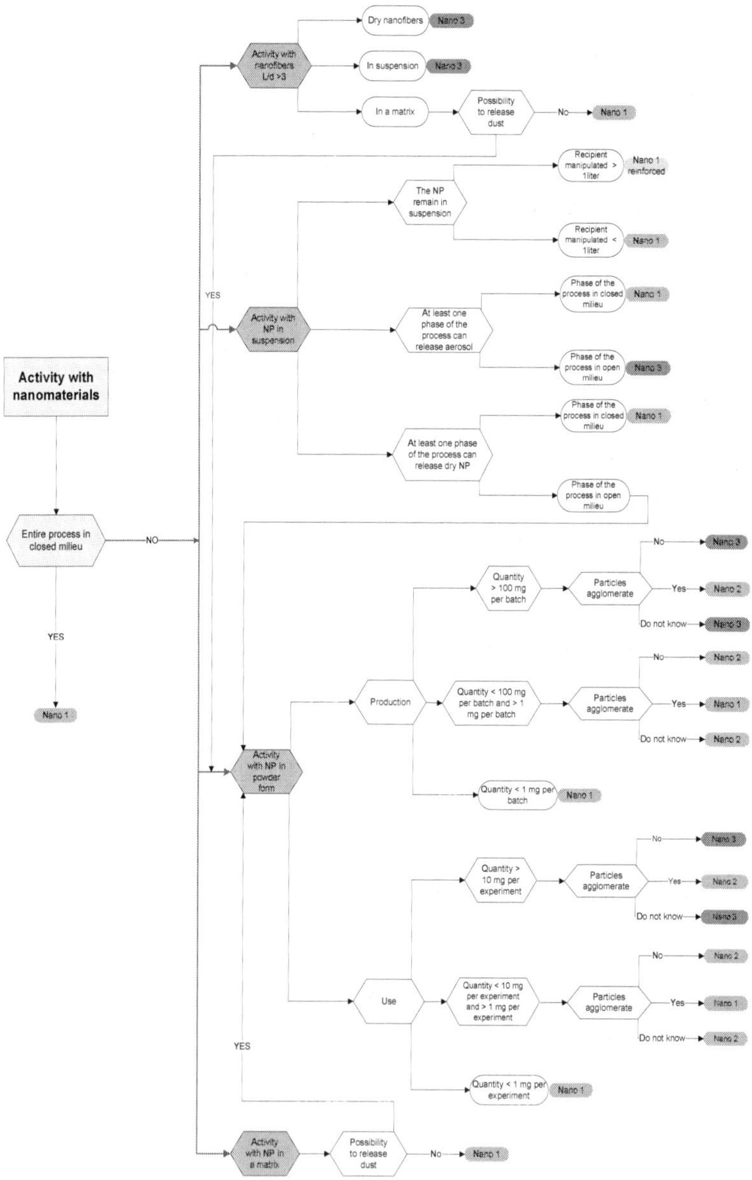

Figure 2: Decisions tree used for determination of Nano hazard type. Questions to be answered by nanomaterials users and producers when determining laboratory (Nano hazard) type. NP: nanoparticles, L/d: length - diameter aspect ratio.

Activity with nanofibers

The scientific community is mostly concerned about the toxicity/ carcinogenicity of manufactured nanofibers (nanomaterials with length-diameter aspect ratio larger than 3) because of their morphological resemblance to asbestos. Inhalation of asbestos fibers is known to induce asbestosis (a progressive fibrotic disease of the lung), lung and pleura cancer. The health hazards of nanofibers are mostly limited to carbon nanotubes and are the subject of an intensive research. Results of already established toxicity studies show a clear morphology-toxicity relationship for carbon nanotubes, [35, 36] as previously observed for asbestos fibers [37]. However, synthesis of nanofibers is being continuously under progress and, as a result, nanofibers can be made out of nearly any material nowadays. Some of them will very likely resemble more closely to asbestos than carbon nanotubes by their size, chemical composition or surface properties. They open the possibility of making nanofibers with undesired harmfulness [38], which could be putatively equal or even higher than the one of asbestos. Hence, all activities, either with dry nanofibers or nanofibers in suspension will situate the laboratory in the *Nano 3* category (Figure 2) except for those where nanofibers are strongly interacting with the matrix (composites), preventing the materials to be released in the environment (refer to activities with nanoobjects in solid matrix).

Activity with Nanoobjects in Powder

The exposure dose is a function of exposure level and duration of exposure [39]. In traditional risk assessment, exposure doses are compared to Threshold Limit Values (TLV). In Switzerland, there are no TLV that were specifically generated for nanomaterials but there are TLV [40] for diesel particles (0.1 mg/m^3), and fumed silica (4 mg/m^3). The British Standards Institute proposed, as a pragmatic guidance, the following [14]: if a material is classified in its larger form as carcinogenic, mutagenic, asthmagenic or a reproductive toxin and a TLV is known, its nano form will have a TLV 10 times smaller. For insoluble materials, their nanoform will have a larger safety margin (1/15 the non-nanoscale TLV); for soluble materials, it is reduced by a half. These considerations are included in our approach.

For nanopowders we distinguished ***production*** and ***handling***(Figure 2). The lower limit up to which a ***production*** laboratory is classified in *Nano 1* is set to 1 mg of nanomaterial present at any given moment. If one assumes a volume of 10 cubic-meters in which particles could spread around an equipment or a person in case of incidents, 1 mg corresponds to the TLV for diesel particles (1 mg/10 m^3 = 0.1 mg/m^3). From a practical perspective, 1 mg constitutes the lower detection limit of many common laboratory balances. Laboratories with more than 1 mg but less than 100 mg are classified as *Nano 2 or Nano 3*, depending on the agglomeration status. Nanomaterials exhibiting large surfaces might be toxic or catalyze the production of toxic substances. Furthermore, nanoparticles often display good transfer into [41] and across epithelial cells [42] and then distribute to other body compartments probably as a function of size and surface properties [43]. Thus, laboratories with single particles or unstable agglomerates are in *Nano 3*. Stable agglomerates and aggregates do not have the nano-specific route of transfer ("normal" transports can still occur) and are expected to affect health more like "classical" ambient air pollution particles [44]. Consequently, activities with agglomerates are classified in *Nano 2*.

Laboratories using more than 100 mg (a considerable quantity in a research environment) are always classified as *Nano 3*.

For the hazard classification of nanoparticles ***handling*** activities, an identical approach as for production activities is taken (Figure 2), with the exception that the upper limit for *Nano 3* is reduced by a factor ten (10 mg). Very often, particles are supplied by other laboratories or external suppliers, where occupational safety and health team cannot control the process as well as for home-made particles. Furthermore, users manipulate such particles more often in confined spaces.

Activity with nanoobjects in suspension

Many applications and investigations use nanoparticles in colloidal suspensions. Nanoparticles in suspension are rarely encountered as bare nanoparticles but have their surface modified in order to ensure colloidal stability or subsequent surface derivatization. This increases the complexity when determining colloids toxic action.

The hazards related to nanomaterials suspension is not only influenced by the nature of particles but also by the dispersant. The decision tree (Figure 2) is organized accordingly: For manipulated quantities superior to 1 liter the nature of the used dispersant (flammable, toxic etc.) is considered equally important: working under the fume hood is mandatory in laboratories classified as "reinforced hazard level Nano 1". If particles remain in suspension and the manipulated quantity is smaller than 1 liter, the laboratory is classified Nano 1 (equivalent to a classical chemical lab).

If aerosols can be released, the equipment has to be placed in a closed environment. Airborne droplets can carry large amounts of nanoparticles into the lungs, and especially small droplets can dry quickly while the solid parts remain airborne. Laboratories with such processes are thus considered as Nano 3.

Activity with nanoobjects in solid matrix

Studying composites with nanoobjects embedded in polymer or in ceramic matrices represents one of the most important activities with nanomaterials at EPFL. The preparation of composites is either treated as "Activity with nanoobjects in suspension" or "Activity with nanoobjects in powder" when performed in solution or in dry conditions, respectively. The laboratory is treated as *Nano 1* if material characterization and post-preparation processing activities do not include any mechanical or thermal treatment. If dust can be released during the manipulation or if composites are friable, laboratory is treated as «Activity with nanoobjects in powder".

Protective Measures

Inhalation and skin contact are considered as most important exposure routes. Measures are organized in consequence. Technical, organizational and personal protective measures for different laboratory (Nano hazard) types are presented in Figure 3, 4 and 5. Even though management of cleaning can be considered as a part of organizational measures, it is separated (Figure 6) to underline its importance.

	Measures		Laboratory		
			Nano 1	Nano 2	Nano 3
Technical	Ventilation	Chemistry lab type (renewal without recycling 5-10 X/h)	x	x	x
		With at least sealed F7 filter (maintenance!) for exiting air			x
		Low pressure in the room	x	x	> 20 mPa
		Capture at source		x	x
	Floor	Flooring	Tiling or linoleum		Resin
	Manipulation under fume hood	Optional	x (1)		
		Compulsory		x	x
	Access restriction	Restricted (magnetic card access control system)		x	x
		Regular lab access control (laboratory key)	x		
		Evidence about exposed people + board to record presence			x
	SAS entrance and exit	Double SAS (if > 100 g ultrafine particles)			x
		Simple SAS (is < 100 g ultrafine particles)		Light SAS	x
		Safety shower			x
	Use of vacuum cleaners	Asbestos type	x	x	x
		Housekeeping type		Forbidden	

Figure 3: Technical safety measures. Technical safety actions applied to laboratory classified 'Nano'. (1) Reinforced type 1 = type 1 plus obligatory manipulation under fume hood.

	Measures		Laboratory Nano 1	Nano 2	Nano 3
Organizational	Restricted access	Authorized persons only		x	x
		Only activities nano in the laboratory			x
	Training	Written working procedures	x	x	x
		Basic training	x	x	x
		Continuous training		x	x
	City/laboratory clothes separation				x
	Conditioning of material contaminated by nano	Toxic (trash bin for toxic)	x	x	x
		Double bag for toxic waste (100 microns thickness)	x	x	x
		Storage of bags in a sealed container	x	x	x
	Elimination of nano substances and products	Liquid waste	Double packaging		
		Solid waste			
	Waste and PPE evacuation	Domestic waste treatment channel	Forbidden		
		Special waste treatment channel	x	x	x
	Transports of "nano-objects"	Simple packaging	Forbidden		
		Double packaging	x	x	x

Figure 4: Organizational safety measures. Organizational safety actions applied to laboratories classified 'nano'.

	Measures		Laboratory Nano 1	Nano 2	Nano 3
Personal	Eyes protection	Safety glasses	x		
		Laboratory mask or close fitting safety goggles		x	x
	Respiratory organs protection	Mask with assisted ventilation			x
		FFP3 mask		x	if < 2 h
	Body protection	Overal with hood - Tyvek style			x
		Non-woven lab coat		x	
		Simple lab coat	x		
		Overshoes		x	x
	Hands protection	2 pairs of adapted gloves		x	x
		1 pair of adapted gloves	x		

Figure 5: Personal safety measures. Personal safety actions applied to laboratories classified 'nano'.

	Measures		Laboratory Nano 1	Nano 2	Nano 3
Cleaning	Who ?	External personnel	x		
		Specially trained external personnel		x	
		Only laboratory personnel			x
	How ?	Wet cleaning only	x	x	x
		Broom	Forbidden		
		Housekeeping type vacuum cleaner	Forbidden		
	Protective equipment	Regular	x		
		The same as for laboratory personnel		x	x
	Supervision	Laboratory responsible		x	x
		Without supervision	x		

Figure 6: Cleaning management. Organization of cleaning for different nano laboratories' types.

Technical measures

As illustrated in Figure 3 laboratories with hazard level *Nano 3* will require rather extensive technical measures with capture at source, exiting air filtering with at least a F7 filter [EN 779 -European Standard for ventilation filters. F7 has 80-90% average efficiency for 0.4 μm particles], and access restrictions using a security vestibule (double door). A double security vestibule with a safety shower is required for each new lab while simple one can be installed for existing laboratories without sufficient space.

Organizational measures

Most organizational protective measures are similar for all laboratory types (see Figure 4). Measures not listed in the figure are the following:

- Each laboratory must have a responsible person (nano-officer).
- An ordering/receiving procedure must be established with identified collecting points.
- Pregnant women are allowed to work with nanomaterials only with a special work authorization issued by an occupational physician.
- Lab safety audits are performed by occupational health and safety specialists.
- Permanent laboratory staff working in *Nano 2* lab and every person working in *Nano 3* are subject to medical surveillance On this question, reports [45] indicate that level of knowledge today doesn't allow proposing a specific medical survey, or indicators of exposure or effects. Still, certain consensus is obtained at international level [46] to recommend that potentially exposed workers should have periodical medical survey with 'conventional' exams, specific for potential target organ. One can think about respiratory tract or cardiovascular system. Results of these medical exams can also be source for data base to make epidemiological studies afterwards.

Personal measures

Personal protective measures (see Figure 5 as well as Figure 7) assign specific equipment's to different hazard levels. As example, a mask with Powered Air Respirator must be used if the work lasts over two hours (*Nano 3*), while P3 (EN 143) or FFP3 (EN 149)/P-100 (USA NIOSH) filter/filtering mask is accepted for shorter work periods. Protection of body parts depends on the hazard level. Two pairs of adapted protective gloves are mandatory when working in *Nano 2* and above.

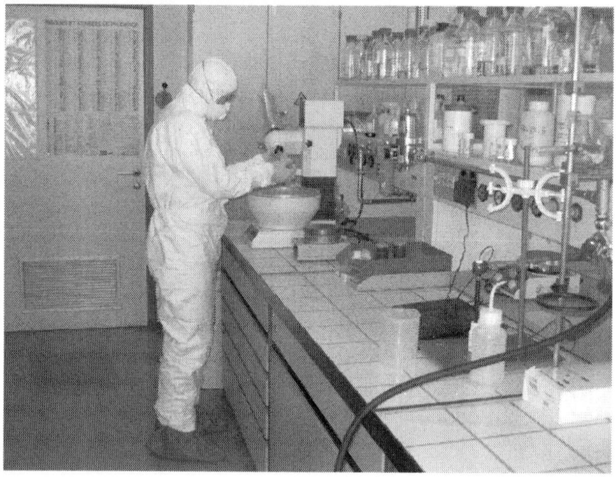

Figure 7: Illustration of personal protective equipment for work with nano. Illustration of personal equipment to be used in a Nano 3 laboratory.

Cleaning management

Only *Nano 1* laboratories can be cleaned by the regular (external) cleaning staff (see Figure 6) wearing protective equipment adapted to work in a chemical laboratory. *Nano 2* must be cleaned by specially trained (external) personnel wearing the same protective equipment as lab employees and under the supervision of the lab responsible. *Nano 3* must be cleaned exclusively by the lab employees themselves wearing the same personal protective equipment as for working and under supervision of a lab responsible. Only in exceptional cases, trained external personnel can be allowed to clean in *Nano 3* laboratories under the supervision of the nano-officer.

CONCLUSIONS

Present knowledge on nanomaterial toxicity is insufficient for completing precise risk assessment. Threshold Limit Values for nanomaterials do not exist nor is there standard equipment for sufficiently detailed routine exposure measurements. However, since preliminary scientific evaluations show that there are reasonable grounds for concern that activity with nanomaterials might have damaging effects on human health; the precautionary principle must be applied. Here we propose practical, clear and simple procedure for

Nano safety and health management, which is a general approach based on the state of the nanomaterial in question (fibers and particles as powder, suspension, or in a solid matrix). New hazard knowledge will be used as it is developed and made available. The procedure proposes pragmatic mitigation measures that laboratories have to take for limiting exposures as much as considered reasonable. The lab responsible is in charge of applying measures adapted to specific activities. The proposed methodology and protective measures are provisional in nature pending the availability of more reliable scientific data. The procedure also allows estimating the number of research groups working in high hazard level laboratories. For reducing investment and operating cost, activities that classify laboratories into the highest hazard level should be centralized as much as possible. The methodology is being implemented at the Ecole Polytechnique de Lausanne in over 100 research labs dealing with nanomaterials; evaluation of its performance will be done when sufficient data are available.

ACKNOWLEDGEMENTS

J. Guzzardi, P. Pugeaud (STI, EPFL), SUVA (non-profit insurance company under Swiss public law), SECO (State Secretariat for Economic Affairs) and Medical services EPFL are gratefully acknowledged for their valuable contribution. This work is partially supported by The Swiss National Science Foundation through the projects numbers 200021-115900 and 205321-125299.

AUTHORS' CONTRIBUTIONS

AG was in charge within the "nanosafe team" of the nano aspects related with physics. She contributed designing the decision tree and the safety measures matrix. She is the paper coordinator and main writer. APF helped designing the decision tree used for determination of laboratory (Nano hazard) type and helped writing the manuscript. She already runs a Nano3 laboratory and gave input on particle and suspension characteristics and characterizations. AM was involved in designing the decision tree used for determination of laboratory (Nano hazard) type and participated in the manuscript writing. He is in charge of a Nano3 laboratory at EPFL since 2003 and he gave input on particles, fibers and composites characteristics and characterizations. MR provided input about toxicological and exposure discussions, control banding knowledge and occupational hygiene strategies.

He also supervised a Master study that investigated the situation of nano-workplaces at EPFL and which helped identify action domains. TM was the initiator and coordinator or the "nanosafe team" whose mission was to analyse nano activities in EPFL labs, to define strategies for laboratories classification and to implement adequate measures. He supervised a Master study on nano activities at EPFL that indentified potential hazards and led to the building of the "nanosafe team". All authors have read and approved the final manuscript.

REFERENCES

1. New Dimensions for Manufacturing: *A UK Strategy for Nanotechnology*. Department of Trade and Industry; 2002.
2. Gupta AK, Gupta M: Synthesis and surface engineering of iron oxide nanoparticles for biomedical applications. *Biomaterials* 2005, 26: 3995–4021. 10.1016/j.biomaterials.2004.10.012
3. Kreyling WG, Semmler M, Moller W: Dosimetry and toxicology of ultrafine particles. *14th International Congress of the International-Society-Aerosols-in-Medicine Jun 14–18; Baltimore, MD* 2003, 140–152.
4. Rothen-Rutishauser BM, Schurch S, Haenni B, Kapp N, Gehr P: Interaction of fine particles and nanoparticles with red blood cells visualized with advanced microscopic techniques. *Environ Sci Technol* 2006, 40: 4353–4359. 10.1021/es0522635
5. ISO/TR 12885: Nanotechnologies -- Health and safety practices in occupational settings relevant to nanotechnologies. International Organization for Standardization; 2008.
6. Dobrovolskaia MA, McNeil SE: Immunological properties of engineered nanomaterials. *Nat Nanotechnol* 2007, 2: 469–478. 10.1038/nnano.2007.223
7. Aitken RJ, Hankin SM, Ross B, Tran CL, Stone S, Fernandes TF, Donaldson K, Duffin R, Chaudhry Q, Wilkins TA, Wilkins SS, Levy LS, Rocks SA, Maynard A: *EMERGNANO: A review of completed and near completed environment, health and safety research on nanomaterials and nanotechnology*. Department for Environment, Food and Rural affairs; 2009.
8. McNeil SE: Nanotechnology for the biologist. *J Leukoc Biol* 2005, 78: 585–594. 10.1189/jlb.0205074
9. Editorial: Lack of progress on nanotoxicology has been highlighted by a tragic accident in China. Nat Nanotechnol 2009, 4: 533. 10.1038/nnano.2009.244
10. Song Y, Li X, Du X: Exposure to nanoparticles is related to pleural effusion, pulmonary fibrosis and granuloma. *Eur Respir J* 2009, 34: 559–567. 10.1183/09031936.00178308
11. New study seeks to link seven cases of occupational lung disease with nanoparticles and nanotechnology[http://go.nature.com/HeLtdg]

12. Nanoparticle exposure and occupational lung disease - six expert perspectives on a new clinical study[http://go.nature.com/XAZkwA]
13. Approaches to Safe Nanotechnology: An Information Exchange with NIOSH Department of Health and Human Services, Centers for Disease Control and Prevention. NIOSH; 2006.
14. Nanotechnologies - Part 2: Guide to safe handling and disposal of manufactured nanomaterials. British Standards Institution; 2007.
15. Meili C, Widmer M, Husmann F, Gehr P, Blank F, Riediker M, Schmid K, Stark W, Limbach L: *Grundlagenbericht zum Aktionsplan «Risikobeurteilung und Risikomanagement synthetischer Nanomaterialien». Umwelt-Wissen Nr. Bundesamt für Umwelt und Bundesamt für Gesundheit*. 2007.
16. Schmid K, Riediker M: Use of nanoparticles in Swiss industry: a targeted survey. *Environ Sci Technol* 2008, 42: 2253–2260. 10.1021/es071818o
17. Nanosafe[http://www.nanosafe.org]
18. Nanosmile[http://www.nanosmile.org/]
19. Gwinn MR, Tran L: Risk Management of nanomaterials. *Nanomed Nanobiotechnol* 2010, 2: 130–137.
20. Murashov V, Engel S, Savolainen K, Fullam B, Lee M, Kearns P: Occupational safety and health in nanotechnology and Organisation for Economic Cooperation and Development. *J Nanopart Res* 2009, 11: 1587–1591. 10.1007/s11051-009-9637-7
21. *Precautionary Matrix for Synthetic Nanomaterials*. Federal Office of Public Health Switzerland; 2008. [http://www.bafu.admin.ch/nanotechnologie]
22. Paik SY, Zalk DM, Swuste P: Application of a pilot control banding tool for risk level assessment and control of nanoparticle exposures. *Ann Occup Hyg* 2008, 52: 419–428. 10.1093/annhyg/men041
23. Asbach C, Kaminski H, Fissan H, Monz C, Dahmann D, Mülhopt S, Paur HR, Kiesling HJ, Herrmann F, Voetz M, Kuhlbusch TAJ: Comparison of four mobility particle sizers with different time resolution for stationary exposure measurements. *J Nanopart Res* 2009, 11: 1593–1609. 10.1007/s11051-009-9679-x
24. Rio Declaration Principle 15 Conference on Environment and Development; Rio de Janeiro 1992.

25. Andorno R: The Precautionary Principle: A New Legal Standard for a Technological Age. *J Int Biotechnol Law* 2004, 1: 11–19. 10.1515/jibl.2004.1.1.11

26. Murashov V, Engel S, Savolainen K, Fullam B, Lee M, Kearns P: Special issue online: Environmental and human exposure of nanomaterials. *J Nanopart Res* 2009.

27. Best Practices Guide to Synthetic Nanoparticle Risk Management. Institut de recherche Robert-Sauvé en santé et en sécurité de travail (IRSST), Canada; 2009.

28. Good Nano Guide[http://www.goodnanoguide.org/Basic]

29. Les Nanomatériaux: Definitions, risques toxicologiques, caractérisation, de l'éxposition professionelle, et mesures de prévention. Institut National de Recherche et de Sécurité; 2008.

30. Best Practices for Handling Nanomaterials in Laboratories[http://ehs.mit.edu/site/sites/default/files/files/University_Best_Practices.pdf]

31. University of New Hampshire: Nanomaterials safety program[http://www.unh.edu/ehs/pdf/UNH-Nanomaterials-Safety-Program.pdf]

32. Balas F, Arruebo M, Urrutia J, Santamaria J: Reported nanosafety practices in reaserch laboratories worldwide. *Nature Nanotechnology*2010, 5: 93–96. 10.1038/nnano.2010.1

33. Questionable safety practices in nanotechnology labs around the world[http://www.nanowerk.com/spotlight/spotid=14778.php]

34. Drezek RA, Tour JM: Is nanotechnology too broad to practise. *Nat Nanotechnol* 2010, 5: 168–169. 10.1038/nnano.2010.37

35. Magrez A, Kasas S, Salicio V, Pasquier N, Seo JW, Celio M, Catsicas S, Schwaller B, Forro L: Cellular toxicity of carbon-based nanomaterials. *Nano Lett* 2006, 6: 1121–1125. 10.1021/nl060162e

36. Poland CA, Duffin R, Kinloch I, Maynard A, Wallace WAH, Seaton A, Stone V, Brown S, MacNee W, Donaldson K: Carbon nanotubes introduced into the abdominal cavity of mice show asbestos-like pathogenicity in a pilot study. *Nat Nanotechnol* 2008, 3: 423–428. 10.1038/nnano.2008.111

37. Hirose A, Nishimura T, Fukumor N, Ogata A, Ohashi N, Kitajima S, Kanno J: Induction of mesothelioma in p53+/- mouse by intraperitoneal application of multi-wall carbon nanotube. *J Toxicol Sci* 2008, 33: 105–116. 10.2131/jts.33.105

38. Magrez A, Horvath L, Smajda R, Salicio V, Pasquier N, Forró L, Schwaller B: Cellular Toxicity of TiO2-Based Nanofilaments. *ACS Nano* 2009, 3: 2274–2280. 10.1021/nn9002067
39. Agency for Toxic Substances and Disease Registry[http://www.atsdr.cdc.gov/HAC/phamanual/toc.html]
40. Valeurs Limites d'exposition aux postes de travail. SUVA, Switzerland; 2009.
41. Rothen-Rutishauser B, Mühlfeld C, Blank F, Musso C, Gehr P: Translocation of particles and inflammatory responses after exposure to fine particles and nanoparticles in an epithelial airway model. *Part Fibre Toxicol* 2007, 4: 1–9. 10.1186/1743-8977-4-9
42. Yacobi NR, DeMaio L, Xie JS, Hamm-Alvarez SF, Borok Z, Kim KJ, Crandall ED: Polystyrene nanoparticle trafficking across alveolar epithelium. *Nanomedicine: NBM* 2008, 4: 139–145.
43. Garcia-Garcia E, Gil S, Andrieux K, Desmaële D, Nicolas V, Georgin D, P Andreux J, Roux F, Couvreur P: A relevant in vitro rat model for the evaluation of blood-brain barrier translocation of nanoparticles. *Cell Mol Life Sci* 2005, 62: 1400–1408. 10.1007/s00018-005-5094-3
44. Brook RD, Franklin B, Cascio W, Hong Y, Howard G, Lipsett M: *Air pollution and cardiovascular disease: a statement for healthcare professionals from the Expert Panel on Population and Prevention Science of the American Heart Association*. 2004.
45. Avis de l'Agence française de sécurité sanitaire de l'environnement et du travail relatif aux nanomatériaux et à la sécurité au travail. AFSSET. 2008.
46. Gaffet E: *Aspects sécurité des nanomatériaux et nanoparticules manufacturés*. Editions Techniques de l'ingénieur; 2009.
47. Sipos B, Duchamp M, Magrez A, Forro L, Barisic N, Kis A, Seo JW, Bieri F, Krumeich F, Nesper R, Patzke GR: Mechanical and electronic properties of vanadium oxide nanotubes. *J Appl Phys* 2009, 105: 074317. 10.1063/1.3103280
48. Smajda R, Andresen JC, Duchamp M, Meunier R, Casimirius S, Hernadi K, Forró l, Magrez A: Synthesis and mechanical properties of carbon nanotubes produced by the water assisted CVD process. *Phys Status Solidi B* 2009, 246: 2457–2460. 10.1002/pssb.200982269

7

Risk Assessment and Risk Management of Nanomaterials in the Workplace: Translating Research to Practice

EILEEN D. KUEMPEL, CHARLES L. GERACI, and PAUL A. SCHULTE

National Institute for Occupational Safety and Health (NIOSH), Education and Information Division, Nanotechnology Research Center, Cincinnati, OH 45226, USA

ABSTRACT

In the last decade since the rise in occupational safety and health (OSH) research focusing on nanomaterials, some progress has been made in generating the health effects and exposure data needed to perform risk assessment and develop risk management guidance. Yet, substantial research gaps remain, as do challenges in the translation of these research findings to OSH guidance and workplace practice. Risk assessment is a process that integrates the hazard, exposure, and dose–response data to characterize risk in a population (e.g. workers), in order to provide health information needed for risk management decision-making. Thus, the research priorities for risk assessment are those studies that will reduce the uncertainty in the key factors that influence the estimates. Current knowledge of OSH in nanotechnology includes the following: (i) nanomaterials can be measured using standard

Citation: EILEEN D. KUEMPEL CHARLES L. GERACI PAUL A. SCHULTE, Risk Assessment and Risk Management of Nanomaterials in the Workplace: The Annals of Occupational Hygiene, Open Access Volume 56, Issue 5, July 2012, Pages 491–505, https://doi.org/10.1093/annhyg/mes040

measurement methods (respirable mass or number concentration), (ii) workplace exposures to nanomaterials can be reduced using engineering controls and personal protective equipment, and (iii) current toxicity testing and risk assessment methods are applicable to nanomaterials. Yet, to ensure protection of workers' health, research is still needed to develop (i) sensitive and quantitative measures of workers' exposure to nanomaterials, (ii) validation methods for exposure controls, and (iii) standardized criteria to categorize hazard data, including better prediction of chronic effects. This article provides a state-of-the-art overview on translating current hazard research data and risk assessment methods for nanomaterials to the development and implementation of effective risk management guidance.

Keywords: hazard assessment, nanomaterials, occupational exposure, occupational exposure limit, occupational health, respirable dust, risk assessment, risk management, hazard assessment

INTRODUCTION

Nanotechnology is a recognized cross-cutting technology that enables applications across all economic sectors [International Organization for Standardization (ISO), 2008; Organization for Economic Cooperation and Development (OECD), 2008]. Yet, the potential adverse health effects remain poorly characterized for many nanomaterials. With the increase in production and use of nanomaterials comes the potential for increased exposure of workers to nanomaterials (Invernizzi, 2011). Guidance on working safely with nanomaterials has been developed in the past decade by government agencies, academia, and occupational health organizations [e.g. The Royal Society, 2004; Maynard and Kuempel, 2005; Oberdorster *et al.*, 2005b; BSI, 2007; ISO, 2009; National Institute for Occupational Safety and Health (NIOSH), 2009a; OECD, 2009; ANSES, 2010; Cornelissen *et al.*, 2011]. However, the extent to which that guidance is followed is not well known, and validation of the effectiveness of the exposure controls and measurement methods for nanomaterials remains a key research need. In the absence of regulatory occupational exposure limits (OELs) for most nanomaterials, a strategy is needed to assess the hazard and determine the appropriate levels of exposure control to protect workers' health.

An effective occupational safety and health(OSH) program for nanomaterials (or hazardous materials generally) integrates the components of basic research, guidance development, and workplace actions (Fig.

1). This integrated scheme can also be viewed as a research-to-practice approach (www.cdc.gov/niosh/r2p/). The current state-of-the-science for nanomaterials with respect to OSH consists of a still-limited but increasing toxicological data base, although there is no standardized framework yet for evaluating and interpreting those data. As such, the translation of research findings to guidance and action has been relatively piecemeal for nanomaterials (as for occupational hazards generally), such that workers may have different levels of health protection depending on the health-effects data available for a specific substance and the extent to which exposure controls are implemented.

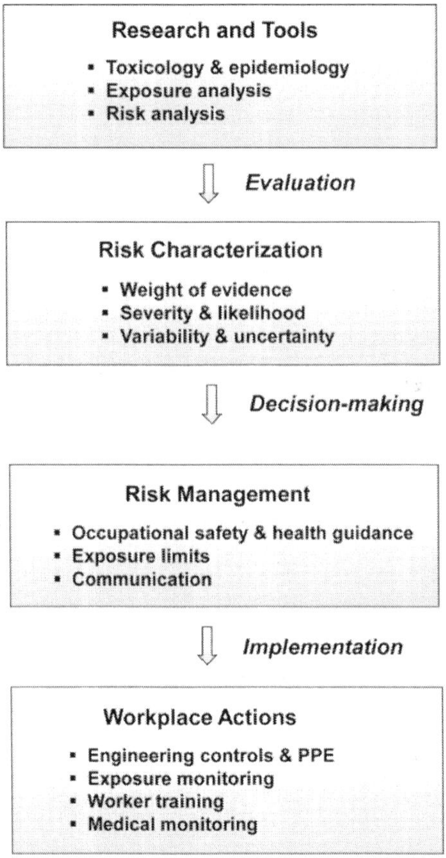

Figure 1: Components for assessing, characterizing, communicating, and managing risks. (PPE: personal protective equipment).

This article provides an overview of what we know and what we still need to know concerning OSH research and practices as it relates to nanomaterials. 'We' refers to the OSH community including the NIOSH Nanotechnology Research Center (NTRC) (NIOSH, 2010a,b). Specific examples [e.g. carbon nanotubes (CNT)] of the data needed for risk assessment to support risk management decision-making, focusing on inhalation hazards, are provided. An integrated process is proposed for the evaluation, decision-making, and implementation of OSH research and guidance.

WHAT WE KNOW

Recent evaluations of current methods for exposure measurement and control have generally shown that these are effective in reducing exposure of workers to nanomaterials (Methner, 2008; NIOSH, 2009a). The next step is to evaluate if these exposure controls are sufficiently health protective, which requires linkage to hazard data and risk estimates. In some cases (e.g. CNT), more sensitive sampling and analytical methods may need to be developed (NIOSH, 2010c). Examples of the state-of-the-science and specific research gaps at each stage of a comprehensive OSH process (Fig. 1) are suggested in Tables 1–3.

Table 1: Examples of research progress and knowledge gaps in predicting hazard potential of nanomaterials[a]

What we know?	What we still need to know?
· Biological responses can depend on particle properties,[b]such as size, shape, surface area, surface chemistry, and solubility	· The difference between the influence of particle properties and other study design differences · Results from assessments with standard materials and response measures
· Some relationships between *in vitro* and 'acute' *in vivo*animal responses have been demonstrated (example: Metal oxides and reactive oxygen species/inflammation responsec	· How to predict 'chronic' responses in animals and humans · A more complete understanding of the role of physical–chemical properties across mode-of-action classes

[a]Examples of physical and biological metrics evaluated in 'Research and tools' in Fig. 1. This list is not comprehensive.

[b]Maynard and Kuempel (2005), Oberdörster *et al.* (2005), OECD (2010b), and Castranova (2011).

[c]Donaldson *et al.* (2010), Rushton *et al.* (2010), and Puzyn *et al.* (2011).

Table 2: Progress and gaps in the information needed to support OSH guidance in nanotechnology workplaces[a]

Where we are?	Where we need to be?
· Have made advances in exposure instrumentation and measurement strategies[b]	· Need sensitive, specific, and quantitative measure of workers' exposure
· Have sensitive measures of biological response in experimental systems[c]	· Need framework to interpret hazard data with respect to workers' health risk
· Have partially delineated the role of nanomaterial physical–chemical properties in toxicity[d]	· Need validated predictive models to make risk management decisions
· Have developed good work practices and general guidance documents[e]	· Need to demonstrate effectiveness of exposure controls in specific work processes

[a]Relates to the evaluation and decision-making steps of a comprehensive OSH process (Fig. 1).

[b]ISO (2007), Brouwer *et al.* (2009), NIOSH (2010c), Bau *et al.* (2010), Görner *et al.* (2010), Johnson *et al.*(2010), and Ramachandran *et al.* (2011).

[c]Sargent *et al.* (2011), Castranova (2011), and Mercer *et al.* (2011).

[d]Duffin *et al.* (2007), Sager and Castranova (2009), Wang *et al.* (2010), and Pauluhn (2011).

[e]BSI (2007), NIOSH (2009a), ISO (2009), OECD (2009), and ANSES (2010).

Table 3: Closing the implementation gap—strategic actions in translating the state-of-the-science to protect nanotechnology workers' health[a]

What we need to do?	How we can get there?
· Enhance risk communication tools	· Develop focused strategies on best practices in exposure control in laboratories, scale-up, and production workplaces
· Emphasize containment and control	· Utilize hazard and control banding tools as a first step and validate effectiveness

· Compare toxicity to existing substances	· Use standardized assays to compare nanoparticle properties and bioactivity with benchmark particles, also look for new effects and/or target organs
· Implement hazard- and risk-based management of nanomaterials	· Develop criteria for nanomaterial hazard categories, occupational exposure controls, and safe and responsible uses of nanomaterials

[a]Examples for implementing risk management guidance to develop effective workplace solutions (final step in a comprehensive OSH process) (Fig. 1).

Nanomaterial OELs

The most important measure to protect workers' health is to minimize (reduce or eliminate) exposures to hazardous substances in the workplace. This is accomplished through effective application of engineering controls and personal protective equipment as part of a risk management program. OELs are intended to guide the control of workplace exposures to levels that would not cause material impairment of health. Relatively few specific OELs have been developed for nanomaterials (reviewed in Schulte *et al.*, 2010; examples in Table 4), and none of these are regulatory standards.

Table 4: Examples of proposed OELs for nanomaterials and the associated exposure control bin

Substance	**OEL (μg m^{-3})a**	**Basis**	**References**	**Exposure control bin (μg m^{-3})**
TiO_2—ultrafine	610b	Estimated human-equivalent concentration to rat subchronic estimated NOAELc of 2 mg m^{-3} (Bermudez *et al.*, 2004), UF of 3	Gamo (2011), Nakanishi (2011a)	100–1 000
TiO_2—ultrafine	300	Working lifetime (45-year) excess risk <1/1,000 (95% LCL estimate) based on benchmark dose model average estimate of particle surface area dose and rat lung tumor response. MOA: secondary genotoxicity from persistent pulmonary inflammation	NIOSH (2011)	
Fullerene (C_{60})	390b	Estimated human-equivalent concentration to rat NOAEL of 0.12 mg m^{-3} (from 28-day inhalation, estimated subchronic equivalent), UF of 9	Shinohara (2011), Nakanishi (2011a)	

MWCNT	50	Estimated human-equivalent rat NOAEL of 0.1 mg m^{-3} (Pauluhn, 2010a) by breathing rate, exposure time, deposition, alveolar macrophage volume, and retention kinetics; no UF	Pauluhn (2010b)	10–100
CNT	30b	Estimated human-equivalent concentration to rat 28-day NOAEL of 0.065 mg m^{-3} for SWCNT and 0.185 for MWCNT. Proposed the lower SWCNT value for all CNTd, UF of 3	Nakanishi (2011a,b)	
CNT and CNF	7 (draft)	Set at LOQ of measurement method. Working lifetime (45 year) excess risk >10% (95% LCL estimate) of early-stage pulmonary inflammation or fibrosis in rat or mouse short-term or subchronic studies	NIOSH (2010c)	1–10
MWCNT	1-2	Adjusted rat NOAEL or LOAEL of 0.1 mg m^{-3}(Pauluhn, 2010a; Ma-Hock *et al.*, 2009, respectively) for exposure day and breathing rate, UF of 25 or 50	Aschberger *et al.* (2010)	

Abbreviations: TiO_2, titanium dioxide; MOA, mode of action; UF, uncertainty factor.

[a]8-h TWA concentration.

[b]OEL (PL): 15-year period-limited OEL. Rat to human adjustments: breathing rate, exposure time, deposition, and body weight.

[c]Bermudez *et al.* (2004) report responses in rat at 2 mg m^{-3}, suggesting that it could be interpreted as a LOAEL.

[d]An OEL (PL) of 0.08 mg m^{-3} was also derived for a MWCNT (44-nm diameter); however, the OEL of 0.03 mg m^{-3} for a SWCNT with the largest specific surface area was proposed as the common value for CNT.

Current proposed nanomaterial OELs are generally low mass concentrations compared with existing OELs for larger respirable particles of the same chemical composition (e.g. metal oxides and carbonaceous particles) (Table 4). Among the proposed nanomaterial OELs, the differences among the same or similar materials may be largely due to different methods and assumptions used to derive the OELs (Table 4) (discussed further below). Despite these differences, there is relative consistency in the OELs within particle types, and these generally fall within the same exposure control bins (e.g. order-of- magnitude categories) (Hewett *et al.*, 2006). Each of these bins is associated with specific engineering control options, based on proven control technologies from other industries (e.g. pharmaceuticals, cosmetics, and dry powder processes) (Naumann *et al.*, 1996). The CNT OEL estimates fall within either 1–10 or 10–100 μg m^{-3} [8-h time-weighted average (TWA)], which may be achievable with containment systems or ventilated enclosures, respectively (Naumann *et al.*, 1996; Ader *et al.*, 2005; Zalk and Nelson, 2008).

The applicability of existing OELs for larger particles to those with nanoscale forms has generally not been evaluated (except, for example, NIOSH, 2011). Moreover, the health basis for existing OELs can vary based on the type of data and methods used to derive the OEL, whether a quantitative risk assessment (QRA) was performed, and the extent to which technical or economic feasibility was given consideration. Thus, standardized risk assessment methods [National Research Council (NRC), 2009; OECD, 2010a] are needed to provide a more consistent health-based approach for setting OELs for nanomaterials.

Risk Assessment: Basic Principles

Occupational health risk assessment is a process to evaluate the hazard, exposure, and dose–response data to characterize risk in workers. Risk estimates provide information to support the development of OELs and other risk management measures. The basic steps in a comprehensive risk assessment (NRC, 2009) include the following:

(i) Problem formulation;
(ii) Risk assessment;
   a. Hazard assessment;
   b. Exposure assessment;
   c. Dose–response assessment;
   d. Risk characterization;
(iii) Risk management and risk communication.

Problem formulation is an initial evaluation of the nature of the hazard, the options for exposure control, and the data needed to distinguish among those options. This step makes risk assessment efficient in providing the information of most value that is needed for decision-making. Hazard assessment is an evaluation of the nature and severity of biological effects (typically in toxicology studies). Exposure assessment involves the measurement or estimation of workers' exposures, by task or full shift. Dose–response assessment (e.g. in animal studies) provides information on the measured or model-estimated dose and the biological responses that are considered relevant to human health. A critical effect level is estimated from the dose–response data (e.g. BMDL, LOAEL, or NOAEL). BMDL is the 95% lower confidence limit estimate on the benchmark dose (maximum likelihood estimate), derived from statistical modeling of the dose–response data. LOAEL is the lowest observed adverse effect level, and NOAEL is the no observed adverse effect level (i.e. the highest dose that is not statistically significantly associated with exposure-attributable adverse effects). The animal critical effect is extrapolated to humans by normalizing the dose across species (e.g. per unit of target tissue) and by adjusting for the dose rate (e.g. assuming 'Haber's principle' that cumulative exposure, i.e. concentration × time, would result in equivalent responses).

QRA is the estimation of the severity and likelihood of an adverse response associated with exposure to a hazardous agent (NRC, 2009). QRA involves not only the best estimate (i.e. central tendency) but also the variability in that estimate, given the data, and the uncertainty in the models and methods used to derive those estimates. Variability is a measure of the distribution of a parameter in a population and can be characterized with measurement data. Uncertainty is the degree of ambiguity, such as in the nature of the hazard or the model used to describe the dose–response relationship. Risk characterization brings together the findings from hazard, exposure, and dose–response assessments to provide information to support the risk management decision-making and risk communication. In the absence of complete risk characterization, or given large uncertainties, additional precaution in the exposure control may be needed to ensure that workers are adequately protected (Schulte and Salamanca-Buentello, 2007).

PROGRESS IN KEY AREAS

Hazard Assessment: Considerations on Nanomaterials

Hazard assessment, in concept, is the same for nanomaterials as for other substances. Current toxicology tests and assays are considered generally applicable to hazard evaluation of nanomaterials (Oberdörster *et al.*, 2005a; OECD, 2008, 2010b). Yet, there may be substance-specific factors that need to be considered in evaluating toxicity, including those for nanomaterials. Some factors that may influence the toxicity of nanomaterials relative to larger particles of the same chemical composition include the following:

- Dose metric;
- Target tissue;
- Physical–chemical properties.

These interrelated factors can affect the uptake and interaction of nanomaterials with biological systems and thus may influence the internal dose and response.

Dose metric

The mass dose metric (exposure concentration in air or lung dose) has been shown to be a poor predictor of toxicity for poorly soluble nanoparticles compared with larger respirable particles (e.g. carbonaceous; metal oxide). Particle surface area dose (Oberdörster *et al.*, 1994; Tran *et al.*, 2000; Bermudez *et al.*, 2002, 2004; Elder *et al.*, 2005; Duffin *et al.*, 2007; NIOSH, 2011) and particle volume (Morrow, 1988; Pauluhn, 2011) have been shown to better predict the lung responses in rats or mice across a range of particle sizes. Agglomeration can influence the deposited dose since the airborne particle size determines the deposition efficiency in the respiratory tract. The form of the nanomaterial to which workers may be exposed should be tested, including if the form is altered by downstream users of the nanomaterial or nanomaterial-containing product, to best estimate the risk of occupational exposure.

Target tissue

The respiratory tract, and specifically the alveolar (gas-exchange) region, is the main target for the deposited dose of respirable particles including nanoparticles. The deposition efficiency of inhaled particles generally increases with decrease in particle size into the nanoscale range [International Commission on Radiological Protection (ICRP), 1994; Maynard and Kuempel, 2005]. Adverse respiratory effects have been reported in workers (exposed to airborne particles or fibers) and in the general population (from exposure to particulate air pollution) (Pope *et al.*, 2002; Rom and Markowitz, 2006). Some types of nanoparticles have been shown to escape normal lung clearance processes (alveolar macrophage phagocytosis) (Renwick *et al.*, 2001) and enter the lung interstitium to a greater extent (Mercer *et al.*, 2010, 2011). Of the nanoparticles studied thus far, the proportion of the mass dose that translocates from the lungs to other organs has been found to be low (Kreying *et al.*, 2002). Yet, nanoparticles may gain access to cells and cell organelles that are not readily accessible to larger particles, and individual nanoparticles have been seen in the cell nucleus, interacting with DNA (Geiser *et al.*, 2005; Sargent *et al.*, 2011b). Ultrafine (nanoscale) particles have also been observed in the mitochondria of treated cells (murine macrophages and BEAS) to a greater extent than fine-size particles; nanoparticles generated more reactive oxygen species per unit mass, causing structural damage (Li *et al.*, 2003). Possible effects outside of the lungs also need to be evaluated, since nanoparticles have been shown to translocate (migrate) from the lungs to the systemic circulation and to other organs in rats and mice (Semmler *et al.*, 2004; Geiser and Kreyling, 2010), as well as from the nasal region via olfactory nerves to the brain in rats (Oberdörster *et al.*, 2004; Elder *et al.*, 2006). These routes have not been demonstrated conclusively in humans, but similar biological structures and mechanisms suggest that these pathways could occur as in animals (Oberdörster *et al.*, 2005).

Physical–chemical properties

Particle size, shape, surface area, surface reactivity, solubility, and functionalization can all influence the particle toxicity (Oberdörster *et al.*, 2005; Castranova, 2011). These properties can influence the internal dose and toxicity at the initial target tissue and distal organs. The toxicity may be either increased or decreased for soluble particles, depending on the biological mode of action (Castranova, 2011; Cho *et al.*, 2012). Because of the greater surface area per unit mass, nanoparticles may be more soluble than larger particles. The degree of agglomeration may also influence the dissolution rate. Some progress has been made toward developing predictive models based on the properties of nanoparticles (e.g. Rushton *et al.*, 2010; Puzyn *et al.*, 2011). These models of quantitative structure–activity relationships (QSAR) require standardized data obtained in controlled experiments in order to evaluate the influence of the specific nanoparticle properties on the dose–response relationships. In the absence of such models, a more time-consuming case-by-case assessment of risk may be necessary. A categorical approach to OEL development (discussed further below) based on the physical–chemical properties could be very useful, in the absence of specific OEL guidance, to make exposure control decisions.

Risk Assessment Methods for Nanomaterials

The current risk assessment process (NRC, 2009) is generally considered to be applicable to nano materials. Although the risk assessment process for nanomaterials is not unique, there are aspects that pertain to risk analyses of sparse or incomplete data. For example, when data are limited, there may be insufficient information to distinguish between alternative plausible models, resulting in large differences (e.g. >10×) in the quantitative estimates (NIOSH, 2010c). In such cases, estimating an OEL band may be an initial step until more precise data can be obtained to develop a more specific OEL.

Recent risk evaluations of nanomaterials (e.g. CNT) have focused on adjustments of the NOAEL or LOAEL from animal short-term or subchronic studies, using various interspecies scaling factors and/or uncertainty factors (Table 4). Differences in methods and assumptions can result in different OELs, even based on the same animal data. Starting from the NOAEL from a subchronic inhalation study of one type of multiwall CNT (MWCNT) (Pauluhn, 2010a), Pauluhn (2010b) estimated an OEL of 50 μg m^{-3} by applying a total interspecies adjustment factor of 2 and no uncertainty factors. Aschberger *et al.* (2010) estimated an OEL of 2 μg m^{-3} based on the same NOAEL but using different interspecies dose scaling and uncertainty factors. Starting from the NOAEL in a 28-day rat intratracheal instillation study of another type of MWCNT, Nakanishi (2011a) derived an OEL of 30 μg m^{-3} as a period-limited (15-year) OEL. None of these assessments fully accounted for the differences in rat and human long-term lung clearance kinetics of inhaled particles generally or for the uncertainty in these estimates for nanoparticles. In a QRA of various types of CNT, NIOSH (2010c) estimated >10% excess risk (95% upper confidence limit estimates) of early-stage lung effects (inflammation, alveolar–interstitial thickening, or fibrosis) over a 45-year working lifetime at 7 μg m^{-3} (8-h TWA), which is the upper limit of quantification (LOQ) for the NIOSH sampling and analytical method for elemental carbon (NIOSH, 2010c). NIOSH set the draft REL at the LOQ and recommended the development of more sensitive measurement methods as a priority research area.

Risk assessment steps

Four standard factors are used to adjust an animal NOAEL (or other effect level) to estimate a human-equivalent dose of inhaled particles (Kuempel *et al.*, 2006):

- Ventilation per exposure day (H/A);
- Deposition fraction (H/A);
- Dose retention kinetics (H/A);

- Interspecies dose normalization (A/H),

where the species-specific factor is for humans (H) or animals (A). These factors are multiplied to obtain the total adjustment factor, which is divided into the animal NOAEL to obtain the estimated human-equivalent NOAEL [e.g. Pauluhn (2010b) or Environmental Protection Agency (EPA) (1994) and similar approaches].

Example of CNT

Quantitative uncertainty

Ventilation and deposition (factors 1 and 2) can be estimated with relatively low uncertainty based on existing measured values in humans and animals and on prediction models for spherical particles that provide estimates of deposition efficiency within the respiratory tract region by airborne particle size, e.g. multiple-path particle dosimetry (MPPD) [CIIT and RIVM, 2006; Applied Research Associates (ARA), 2011]. In contrast, retention kinetics and dose normalization (factors 3 and 4) can have a large influence on the risk estimates due to the relatively limited data available to evaluate alternative models and assumptions in the estimation of the human-equivalent lung dose. Lack of adjustment for interspecies differences in lung clearance kinetics (factor 3) has the same effect as assuming simple steady-state kinetics at the same rate in both species. This is clearly incorrect based on available data of particle clearance in animals and humans, e.g. Snipes (1989) estimated a 10× slower long-term clearance rate in humans than in rats. Moreover, the ICRP (1994) clearance model-based estimates of the human-equivalent chronic lung burdens (45-year working lifetime) are a factor of ~30× greater than the estimated equivalent rat lung burden after chronic (2-year) exposure [estimated at 0.1 mg m^{-3}, NOAEL in Pauluhn (2010a), in MPPD rat and human models (CIIT and RIVM, 2006; NIOSH, 2010c; ARA, 2011)]. A recent human respiratory tract model update would increase the average human-retained lung dose estimates by another factor of 2–3 (Gregoratto *et al.*, 2010, 2011). Thus, assumptions about dose rate and clearance kinetics can have a substantial influence (~10–100×) on the estimate of the human-equivalent lung dose of poorly soluble particles, and therefore on the OEL estimate.

Differences in the interspecies dose normalization assumptions (factor 4) can also have a moderately large influence on the human-equivalent dose estimates. For example, normalizing the dose by the average alveolar epithelial surface area (0.4 m^2/102 m^2) (rat/human) (Mercer *et al.,* 1994) versus by the total alveolar macrophage cell volume (3.0×10^{10} μm^3/3.5×10^{13} μm^3) (rat/human) (Pauluhn, 2010b) results in a ~4.5× difference in the human-equivalent dose estimate.

Given the uncertainty in the CNT lung dose estimates in workers, the NIOSH risk assessment evaluated the bounds of the estimated lung doses, by assuming either normal clearance or no clearance of the deposited dose predicted from spherical particle models (CIIT and RIVM, 2006; NIOSH, 2010c). Some evidence suggests that the true lung dose estimate may lie within these bounds, i.e. CNT particle clearance was observed to be slower than expected for other respirable poorly soluble particles at low airborne mass concentrations (Pauluhn, 2010a). Risk estimate differences due to lung dose assumptions were greater (~4–5×) than those due to the interstudy differences (~2×) (Table 5). The OELs are all based on short-term or subchronic data, and none explicitly address possible carcinogenic end points (Table 4).

Table 5: Human-equivalent benchmark concentration estimates for multiwall CNT, associated with estimated 10% excess risk in rat subchronic inhalation studies (NIOSH, 2010c)

Study	Reponse	Working lifetime 8-h TWA ($\mu g\ m^{-3}$)
Deposited lung dose		
MaHock *et al.* (2009)	Granulomatous inflammation	0.48 (0.19)
Pauluhn (2010a)	Alveolar–interstitial thickening	0.8 (0.41)
Retained lung dose		
MaHock *et al.* (2009)	Granulomatous inflammation	2.7 (1.0)
Pauluhn (2010a)	Alveolar–interstitial thickening	3.5 (1.6)

Note: The upper LOQ of analytical method to measure elemental carbon is 7 $\mu g\ m^{-3}$ (NIOSH Method 5040).

*Maximum likelihood estimate and 95% LCL.

Qualitative uncertainty

The structural similarities of CNT that resemble asbestos have lead to concerns about asbestos-type pathology (Takagi *et al.*, 2008; Jaurand *et al.*, 2009; Donaldson *et al.*, 2010). In addition to the noncancer lung effects (pulmonary inflammation and fibrosis) observed in rats and mice exposed to single-walled CNTs (SWCNTs) and MWCNTs (NIOSH, 2010c), some types of MWCNTs have been shown to induce mesothelioma in rats (Fischer 344/Brown Norway F1 hybrids) by intraperitoneal injection (IP) (either 0.5 or 5 mg MWCNT/rat, twice with a 1-week interval) (Nagai *et al.*, 2011). In contrast, cancer was not observed in Wistar rats 24 months after IP administration (2 or 20 mg/rat) of short MWCNT (<1 μm), although a clear carcinogenic response (~35% vs. ~4% in vehicle control) was observed at a 2 mg/rat dose of UICC crocidolite (Muller *et al.*, 2009). Likewise, pulmonary inflammation was observed in mice following IP of long (5–20 μm), thin, rigid CNT, but not of short or tangled CNT (<1 μm), suggesting a biological response consistent with the fiber paradigm (Poland *et al.*, 2008; Murphy *et al.*, 2011; Osmond-McLeod *et al.*, 2011).

Recent *in vitro* studies have observed specific genotoxic effects and associated biological mechanisms from certain MWCNTs and SWCNTs (Sargent *et al.*, 2009; Sargent *et al.*, 2011a,b). Significant spindle disruption and chromosome abnormality (aneuploidy) were observed at a low mass dose (0.024 μg SWCNTs cm^{-2} cells). Moreover, the SWCNTs were seen to be integrated with the DNA and the microtubule structures (Sargent *et al.*, 2011b). A high death rate occurred initially after the exposure to SWCNTs in both immortalized normal human bronchial epithelial cells (BEAS-2B) and primary normal human small airway epithelial cells (SAEC), followed by an increase in cell proliferation rates, which suggests an increased probability of passing on the genetic damage (aneuploidy) to daughter cells. Aneuploidy is a key event in the progression of some cancers (Kops *et al.*, 2005), and similar biological mechanisms have been observed with chrysotile and crocidolite asbestos (Cortez *et al.*, 2011; Yegles *et al.*, 1995). In addition, micronuclei (chromosome fragments that are not incorporated into the nucleus at cell division) were observed *in vitro* in cells (SAEC) that were incubated with carbon nanofibers (CNF), crocidolite asbestos, or SWCNTs (Kisin *et al.*, 2011). Micronucleated pneumocytes were also observed in rats after a single intratracheal dose of MWCNTs (Muller *et al.*, 2008). K-ras gene mutations, which have been reported in some lung tumors in humans and mice (Jackson *et al.*, 2006), were observed in mouse lungs following inhalation exposure to SWCNTs (Shvedova *et al.*, 2008).

These studies provide compelling evidence that at least some forms of CNT are potentially carcinogenic to humans. Typically, carcinogen classifications by authoritative bodies (e.g. International Agency for Cancer) have been based on human evidence or on chronic bioassays done in rats or mice (e.g. US National Toxicology Program, NTP). A key question for nanotechnology (and OSH generally) is how the findings from the biomechanistic studies should be interpreted in cancer classification decision-making. In view of the CNT evidence, a recent paper asks what specific actions should be taken now to protect workers before chronic animal bioassay data become available (Schulte *et al.*, 2012). Clearly, a high level of exposure containment and control would be prudent, especially for the thin, rigid CNT structures. Development and validation of a QSAR model to predict cancer potential based on CNT structure is a key research need; however, standardized, quantitative dose–response data are generally not yet available.

In addition to CNT, other chemical forms (e.g. metal compounds) and fiber-like (high aspect ratio) structures (e.g. nano rods and wires) are being developed for an array of nanotechnology applications. Understanding the properties influencing the cancer potential may facilitate the development of safer nanomaterials by design (Donaldson *et al.*, 2010).

Exposure Measurement and Control

Some of the NIOSH NTRC field research team studies have reported relatively large reductions in the airborne particle exposures after installation of standard engineering controls (NIOSH, 2010a). For example, during nanometal oxide reactor cleanout, the average percent reduction in airborne particulate was 96(±6)% based on particle counts or 88(±12)% based on particle mass by use of local exhaust ventilation (Methner, 2008). In a laboratory case study, the use of benchtop enclosures prevented the release of carbon nanoparticles during sonication (Johnson *et al.*, 2010). The enclosure was placed on a ventilated benchtop (100 ft min^{-1}). Before installation of exposure controls, airborne MWCNT bundles were observed by transition electron microscope (TEM), and none were detected in the samples collected after the enclosure was installed (Johnson *et al.*, 2010).

A study of laboratory fume hoods showed that the hood design affects the nanoparticle release, and an air-curtain hood design (with a different airflow pattern) significantly reduced workers' exposures (Tsai *et al.*, 2010). A study of the filtration performance of a NIOSH N95 respirator showed that it meets the NIOSH respirator certification criteria (>95% filtration efficiency), although the most penetrating particle size was 50 nm in diameter (~2% filter penetration) (Rengasamy and Eimer, 2011).

Investigation of Nanotechnology Workers' Health

NIOSH and others recommend evaluation of the need for medical monitoring of nanotechnology workers, as well as consideration of exposure registries, and development of epidemiological studies (Nasterlack *et al.*, 2008; Schulte *et al.*, 2008; NIOSH, 2009a,b; Schulte and Trout, 2011; Boutou-Kempf, 2012). NIOSH recently began a study of US workers exposed to engineered carbon nanoparticles (Dahm *et al.*, 2011; Schubauer-Berigan *et al.*, 2011), which included identification of the eligible companies and the most feasible population for study, followed by characterization of workers' exposures to CNT and CNF in manufacturing and distribution. These exposure data include respirable particle mass, number, and active surface area; personal full-shift daily exposure; and targeted sampling of the tasks associated with the highest exposures. Among the 30 companies (of the 61 eligible) that completed a questionnaire, the workforce size was >620, with 2–100 workers per company. Most (60%) of the operations were full scale, and another ~20% were planning to scale-up within 5 years. The materials produced and used included ~70% CNT and ~30% graphene, fullerenes, or carbon or polymer nanofibers. Assessing the health of nanomaterial workers is a critical component of responsible development of the technology (Schulte *et al.*, 2008; Schulte and Trout, 2011; Schulte *et al.*, 2012).

STRATEGIC GOALS: WHAT WE STILL NEED TO KNOW

Key questions concerning working with nanomaterials include the following:

- Are workers being protected?
- Are we effectively translating the research findings into workplace practice?
- Are occupational health guidance and standards keeping pace with the development of new nanomaterials?
- Are we effectively communicating the health risks and protective measures?

To the extent that the current OSH guidance is being followed, it would be expected to result in reduction or elimination of workers' exposures to engineered nanoparticles and potential adverse health effects. Thus, exposures of workers (e.g. to airborne asbestos fibers in textile manufacturing) that have occurred historically would need to be prevented in the production and use of nanomaterials (e.g. CNTs that are used to make high-strength fibers) (Fig. 3). Yet, to fully realize OSH goals for nanotechnology workers, more specific and verified information and guidance (Table 2), as well as measures to implement exposure control solutions in the workplace (Table 3), is needed. Development of more specific guidance on exposure control strategies (e.g. by process and task) is an example of translating research to practice (e.g. Methner *et al.*, 2008; NIOSH, 2010a; Tsai *et al.*, 2010). Such information may be especially helpful to workers and employers in research laboratories and small pilot operations that may not have dedicated OSH programs. Control banding strategies provide a useful decision logic for the initial selection of exposure controls (Paik *et al.*, 2008; ANSES, 2010). Validation of the effectiveness of these exposure control recommendations, and refinement as needed, is also an essential step.

Figure 3: Comparison of spinning operations for asbestos or CNT. (3a) Asbestos thread-making machine with spools of asbestos thread (c. 1930–1960). [Source: Public Health Image Library. Available at: phil.cdc.gov/phil/home.asp (ID#:9646)] (3b) Spinning SWCNT into high-strength 'super rope' fibers (early 2000s) [Source: Ericson et al. (2004). Reprinted with permission from American Association for the Advancement of Science (AAAS)].

Research Needs for Nanomaterials Risk Assessment

The value of information in risk assessment depends largely on the extent to which it reduces uncertainty in the risk estimates (Fig. 2). If a parameter does not have a large influence on the risk estimates, then a higher degree of uncertainty may be acceptable. However, if a risk estimate is highly dependent on a given assumption, then a greater level of effort or research priority in that area would be of value by improving the utility of the risk characterization.

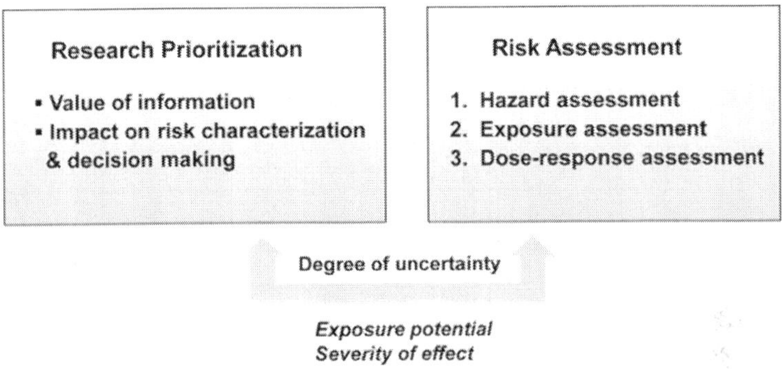

Fig. 2: Focused interaction between risk assessment and research needs.

Cross-cutting research needs

Risk assessment models and methods for various types of inhaled particles share some of the same steps, including (i) lung dose estimation, (ii) interspecies normalization (scaling) of dose, and (iii) temporal extrapolation of dose and response. Research and standardized approaches to reduce the uncertainty in these estimates would be useful for estimating risk and deriving exposure limits for CNTs and other nanomaterials.

Standardized testing

To facilitate comparison of results across studies and particle types, standardized assays and response endpoints are needed. Standardized assays and tiered testing approaches have been recommended (Oberdorster *et al.*, 2005; OECD, 2008, 2010b), yet a relatively small number of the toxicology studies of nanomaterials provide sufficient data for risk assessment, and only a very few studies provide comparable dose–response data that can be evaluated across studies and particle types (NIOSH, 2010c).

Evaluation and validation of these assays for nanomaterials are also necessary. For example, for CNT, the fibrotic response may not be well predicted by pulmonary inflammation, as it may resolve at dose levels that nonetheless are associated with persistent alveolar septal thickening and fibrosis (Shvedova *et al.*, 2005, 2008; Pauluhn, 2010; Mercer *et al.*, 2011). The fibrotic mechanism may be related to the very thin, long CNT structures that mimic the epithelial basement membrane and stimulate fibrotic cell growth (Wang *et al.*, 2010).

Exposure measurement and control

Two challenges for future research include (i) demonstration of the effectiveness of reapplying proven control technologies from other industries (such as pharmaceuticals, cosmetics, and dry powder detergents) that are capable of achieving a high level of exposure control (e.g. in the microgram per cubic meter range), and (ii) until more sensitive and specific analytical methods are developed, background contributions must be factored into any nanomaterial exposure assessment strategy developed by a professional industrial hygienist. Effective risk communication is essential to translate research to risk management practice.

Categorical Approach to OELs

Given the vast number of substances that require testing, alternative test methods are needed to increase the database for evaluating the hazard and determine the level of exposure control needed. Specifically, data that permit selection among the exposure control options are needed. For example, *in vitro* and cell free test methods have been developed using assays of cytotoxicity or reactive oxygen generation (Rushton *et al.*, 2010; Donaldson *et al.*, 2008). These current assays are useful for initial screening and priority setting for subsequent testing, although further evaluation of the association between *in vitro* and *in vivo* responses is needed before *in vitro* assays could fully replace standard *in vivo* assays (Landseidel *et al.*, 2010). A combination approach may also be feasible, by using existing animal dose–response data for well-characterized benchmark particles (for which risk has been quantified), along with short-term *in vivo* or *in vitro* assays of an array of nanomaterials with similar chemical–physical properties and biological mode of action (e.g. carbonaceous particles; metal oxides) (Kuempel *et al.*, 2007, 2011). The comparative toxicity data could then be used in a parallelogram type of analysis (Schoeny and Margosches, 1989; Sobels, 1993) to infer the risk by the nanomaterial in comparison to the benchmark particle. Development and evaluation of categorical approaches to support OEL development is a priority area of the NIOSH NTRC strategic plan (NIOSH, 2010a).

CONCLUDING REMARKS

Although there is uncertainty in the risk assessment of nanomaterials, it can be characterized to some extent. An evaluation of the main steps in the risk assessment process for inhaled particles shows which steps and assumptions have the largest influence on the risk estimates. This evaluation identifies the type of study data needed to reduce the uncertainty and the research priorities needed to obtain those data. Some of these steps involve uncertainty that is specific to the nanomaterial (e.g. CNT lung clearance), and some are more broadly applicable (e.g. extrapolating animal dose to humans; role of dose rate on the adverse effect). Other information is relatively well known (e.g. deposition fraction of inhaled particles in the respiratory tract regions, given the breathing parameters and the airborne particle size). Targeted research using standardized test methods and response endpoints would facilitate comparative toxicity assays and reduce uncertainty in risk assessment across nanomaterials. In addition to inhalation exposure (the focus of this article), dermal and other potential routes of exposure to nanomaterials in the workplace should be evaluated, as well as other possible effects beyond the lungs (NIOSH, 2010a).

Considerable variability exists in the types of nanomaterials, including in the chemical composition, structure, and functionalization. Yet there are relatively few options for exposure control. Determining in which 'bin' a nanomaterial fits could be difficult with relatively little data. For example, additional information may be obtained through comparative toxicity to benchmark (reference or control) particles. Such a strategy may also facilitate comparison across particle types and development of safer nanomaterials. A key challenge, in order for nanotechnology to deliver on its promise of societal benefit, is to ensure that protection of workers' health is being met. Implementing effective measures to reduce or eliminate occupational exposures is an early step in a responsible life-cycle approach and the primary approach needed to prevent adverse health effects in workers producing or using nanomaterials.

ACKNOWLEDGEMENTS

This article is based in part on a keynote presentation at the Institut National de Recherche et de Sécurité (INRS) Occupational Health Research Conference: Risks Associated to Nanoparticles and Nanomaterials, by E.D.K., Ph.D., on 5 April 2011, in Nancy, France. Contributions to that presentation by NIOSH colleagues are gratefully acknowledged: P.A.S., Ph.D.; C.L.G., Ph.D.; Vincent Castranova, Ph.D.; Linda Sargent, Ph.D.; Elena Kisin, Ph.D.; Dale Porter, Ph.D.; Mary Schubauer-Berigan, Ph.D.; Matt Dahm, M.S.; Samy Rengasamy, Ph.D.; Ken Martinez, M.S.; and Mark Methner, Ph.D. We would also like to thank Dr. Linda Sargent for providing additional information on the results of her recently published studies.

REFERENCES

1. ANSES 2010 Development of a specific control banding tool for nanomaterials Maisons-Alfort Cedex, France Agence nationale de sécurité sanitare
2. ARA 2011 Multiple-path particle deposition (MPPD 2.1): a model for human and rat airway particle dosimetry Raleigh, NC Applied Research Associates, Inc
3. Aschberger K Johnston HJ Stone V et al. 2010 Review of carbon nanotubes toxicity and exposure—appraisal of human health risk assessment based on open literature Crit Rev Toxicol 40 759 90
4. Bau S Witschger O Gensdarmes F et al. 2010 A TEM-based method as an alternative to the BET method for measuring off-line the specific surface area of nanoaerosols Powder Technol 200 190 201
5. Bermudez E Mangum JB Asgharian B et al. 2002 Long-term pulmonary responses of three laboratory rodent species to subchronic inhalation of pigmentary titanium dioxide particles Toxicol Sci 70 86 97
6. Bermudez E Mangum JB Wong BA et al. 2004 Pulmonary responses of mice, rats, and hamsters to subchronic inhalation of ultrafine titanium dioxide particles Toxicol Sci 77 347 57
7. Boutou-Kempf O 2012 Department of Occupational Health (DST), French Institute of Public Health Surveillance (InVS). Occupational Health Feasibility elements for epidemiological surveillance of workers exposed to intentionally produced nanomaterials Institut de Veille Sanitaire. [French Institute for Public Health Surveillance]
8. Brouwer D van Duuren-Stuurman B Berges M et al. 2009 From workplace air measurement results toward estimates of exposure? Development of a strategy to assess exposure to manufactured nano-objects J Nanopart Res 11 1867 81
9. BSI 2007 Nanotechnologies—Part 2: guide to safe handling and disposal of manufactured nanomaterials. PD 6699-2:2008 London BSI Group
10. Castranova V 2011 Overview of current toxicological knowledge of engineered nanoparticles JOEM 53 Suppl. 6 S14 7
11. Cho WS Duffin R Thielbeer F et al. 2012 Zeta potential and solubility to toxic ions as mechanisms of lung inflammation caused by metal/metal-oxide nanoparticles Toxicol Sci 12 January 2012 [Epub ahead of print]

12. Cornelissen R Fongeneelen F van Broekhuizen P et al. 2011 Guidance working safely with nanomaterials and nanoproducts: a guide for employers and employees. Version 1.0 – May 2011 Dutch Social Partners FNV, VNO-NCV and CNV
13. CIIT, RIVM 2006 Multiple-path particle dosimetry (MPPD, version 2.0): a model for human and rat airway particle dosimetry Research Triangle Park, NC Centers for Health Research (CIIT) the Netherlands National Institute for Public Health and the Environment (RIVM)
14. Dahm MM Evans DE Schubauer-Berigan MK et al. 2011 Occupational exposure assessment in carbon nanotube and nanofiber primary and secondary manufacturers Ann Occup Hyg. 8 December [Epub ahead of print]
15. de Araujo Cortez B Quassollo G Caceres A et al. 2011 The fate of chrysotile-induced multipolar mitosis and aneuploid population in cultured lung cancer cells PLoS One 6 e18600 1 13
16. Donaldson K Borm PJ Oberdorster G et al. 2008 Concordance between in vitro and in vivo dosimetry in the proinflammatory effects of low-toxicity, low-solubility particles: the key role of the proximal alveolar region Inhal Toxicol 20 53 62
17. Donaldson K Murphy F Schinwald A et al. 2011 Identifying the pulmonary hazard of high aspect ratio nanoparticles to enable their safety-by-design Nanomedicine (Lond) 6 143 56
18. Donaldson K Murphy FA Duffin R et al. 2010 Asbestos, carbon nanotubes and the pleural mesothelium: a review of the hypothesis regarding the role of long fibre retention in the parietal pleura, inflammation and mesothelioma Part Fibre Toxicol 7 5
19. Ericson LM Fan H Peng H et al. 2004 Macroscopic, neat, single-walled carbon nanotube fibers Science 305 1447 50
20. Elder A Gelein R Finkelstein JN et al. 2005 Effects of subchronically inhaled carbon black in three species. I. Retention kinetics, lung inflammation, and histopathology Toxicol Sci 88 614 29
21. Elder A Gelein R Silva V et al. 2006 Translocation of inhaled ultrafine manganese oxide particles to the central nervous system Environ Health Perspect 114 1172 8
22. EPA 1994 Methods for derivation of inhalation reference concentrations and application of inhalation dosimetry Research Triangle Park, NC U.S. Environmental Protection Agency. EPA report no. EPA/600/8-

90/066F

23. Gamo M 2011 Risk assessment of manufactured nanomaterisl: titanium dioxide (TiO2. Final report issued on 22 July 2011. New Energy and Industrial Technology Development Organization (NEDO) project (P06041) "Research and Development of Nanoparticle Characterization Methods." National Institute of Advanced Industrial Science and Technology (AIST). Available at http://www.aist-riss.jp/main/?ml_lang=en
24. Geiser M Rothen-Rutishauser B Kapp N et al. 2005 Ultrafine particles cross cellular membranes by nonphagocytic mechanisms in lungs and in cultured cells Environ Health Perspectives 113 1555 60
25. Geiser M Kreyling WG 2010 Deposition and biokinetics of inhaled nanoparticles Part Fibre Toxicol 7 2
26. Görner P Simon X Wrobel R et al. 2010 Laboratory study of selected personal inhalable aerosol samplers Ann Occup Hyg 54 165 87
27. Gregoratto D Bailey MR Marsh JW 2010 Modelling particle retention in the alveolar-interstitial region of the human lungs J Radiol Prot 30 491 512
28. Gregoratto D Bailey MR Marsh JW 2011 Particle clearance in the alveolar-interstitial region of the human lungs: model validation Radiat Prot Dosimetry 144 353 6
29. Hewett P Logan P Mulhausen J et al. 2006 Rating exposure control using Bayesian decision analysis J Occup Environ Hyg 3 568 81
30. ICRP 1994 Human respiratory tract model for radiological protection. ICRP publication no. 66. In: Smith H Annals of the ICRP Tarrytown, NY International Commission on Radiological Protection
31. Invernizzi N 2011 Nanotechnology between the lab and the shop floor: what are the effects on labor? J Nanopart Res 13 2249 68
32. ISO 2007 Workplace atmospheres—ultrafine nanoparticle and nano-structured aerosols—inhalation, exposure characterization and assessment. Document No. ISO/TR 27628:2007 Geneva International Organization for Standardization
33. ISO 2008 Nanotechnologies — health and safety practices in occupational settings relevant to nanotechnologies. Document No. ISO/TR ISO/TR 12885:2008(E) Geneva International Organization for Standardization
34. ISO 2009 ISO/TC 229 Nanotechnologies working group 3– health,

safety and the environment Project Group 6, "Guide to safe handling and disposal of manufactured nanomaterials", Draft Report, 9 June 2009, Seattle, Washington, U.S.A. NANO TC229 WG 3/PG 6 012–2009 Geneva International Organization for Standardization

35. Jackson MA Lea I Rashid A et al. 2006 Genetic alterations in cancer knowledge system: analysis of gene mutations in mouse and human liver and lung tumors Toxicol Sci 90 400 18
36. Jaurand MC Renier A Daubriac J 2009 Mesothelioma: Do asbestos and carbon nanotubes pose the same health risk? Part Fibre Toxicol 6 16
37. Jiang J Oberdörster G Elder A et al. 2008 Dose nanoparticle activity depend upon size and crystal phase? Nanotoxicology 2 33 42
38. Johnson DR Methner MM Kennedy AJ et al. 2010 Potential for occupational exposure to engineered carbon-based nanomaterials in environmental laboratory studies Environ Health Perspect 118 49 54
39. Kisin E Murray AR Sargent L et al. 2011 Genotoxicity of carbon nanofibers: are they potentially more or less dangerous than carbon nanotubes or asbestos? Toxicol Appl Pharmacol 252 1 10
40. Kops GJ Weaver BA Cleveland DW 2005 On the road to cancer: aneuploidy and the mitotic checkpoint Nat Rev Cancer 5 773 85
41. Kreyling WG Semmler M Erbe F et al. 2002 Translocation of ultrafine insoluble iridium particles from lung epithelium to extrapulmonary organs is size dependent but very low J Toxicol Environ Health 65 1513 30
42. Kuempel ED Tran CL Castranova V et al. 2006 Lung dosimetry and risk assessment of nanoparticles: evaluating and extending current models in rats and humans Inhal Toxicol 18 717 24
43. Kuempel ED Geraci CL Schulte PA 2007 Risk assessment approaches and research needs for nanoparticles: an examination of data and information from current studies. Proceedings of the NATO Advanced Research Workshop on Nanotechnologoy: Toxicological Issues and Environmental Safey Varna, Bulgaria 12–17 August 2006 Simeonova P Opopol N Luster M Nanotechnology: toxicological issues and environmental safety Springer-Verlag, NY 119 45
44. Kuempel ED Geraci CL Castranova V et al. 2011 Development of Risk-based Nanomaterial Groups for Occupational Exposure Control. Presented at the 5th International Conference on Nanotechnology – Occupational and Environmental Health (NanOEH, 10 August 2011

Boston, MA. [Submitted to J Nanoparticle Research]

45. Landsiedel R Ma-Hock L Kroll A et al. 2010 Testing metal-oxide nanomaterials for human safety Ad Mat 22 2601 27
46. Li N Sioutas C Cho A Schmitz D et al. 2003 Ultrafine particulate pollutants induce oxidative stress and mitochondrial damage Environ Health Perspect 111 455 60
47. Ma-Hock L Treumann S Strauss V et al. 2009 Inhalation toxicity of multi-wall carbon nanotubes in rats exposed for 3 months Toxicol Sci 112 468 81
48. Maynard AM Kuempel ED 2005 Airborne nanostructured particles and occupational health J Nanopart Res 7 587 614
49. Mercer RR Russell ML Roggli VL et al. 1994 Cell number and distribution in human and rat airways Am J Respir Cell Mol Biol 10 613 24
50. Mercer RR Hubbs AF Scabilloni JF et al. 2010 Distribution and persistence of pleural penetrations by multi-walled carbon nanotubes Part Fibre Toxicol 7 1 11
51. Mercer RR Hubbs AF Scabilloni JF et al. 2011 Pulmonary fibrotic response to aspiration of multi-walled carbon nanotubes Part Fibre Toxicol 8 21
52. Methner M 2008 Engineering case reports: Effectiveness of local exhaust ventilation in controlling engineered nanomaterial emissions during reactor cleanout operations J Occup Environ Hyg 5 D63 9
53. Morrow PE 1988 Possible mechanisms to explain dust overloading of the lungs Fund Appl Toxicol 10 369 84
54. Muller J Decordier I Hoet PH et al. 2008 Clastogenic and aneugenic effects of multi-wall carbon nanotubes in epithelial cells Carcinogenesis 29 427 33
55. Muller J Delos M Panin N et al. 2009 Absence of carcinogenic response to multiwall carbon nanotubes in a 2-year bioassay in the peritoneal cavity of the rat Toxicol Sci 110 442 8
56. Murphy FA Poland Ca Duffin R et al. 2011 Length-dependent retention of carbon nanotubes in the pleural space of mice initiates sustained inflammation and progressive fibrosis on the parietal pleura Am J Pathol 178 2587 99
57. Nagai H Okazaki Y Chew SH et al. 2011 Diameter and rigidity of multiwalled carbon nanotubes are critical factors in mesothelial injury

and carcinogenesis Proc Natl Acad Sci U S A 108 1 9

58. Nakanishi J 2011 Risk Assessment of Manufactured Nanomaterials "Approaches" - Overview of approaches and Results - Final report issued on 17 August 2011. New Energy and Industrial Technology Development Organization (NEDO) project (P06041) "Research and Development of Nanoparticle Characterization Methods. National Institute of Advanced Industrial Science and Technology (AIST). Available at http://www.aist-riss.jp/main/?ml_lang=en
59. Nakanishi J 2011 Risk Assessment of Manufactured Nanomaterials: Carbon Nanotubes (CNT). Final report issued on 12 August 2011. New Energy and Industrial Technology Development Organization (NEDO) project (P06041) "Research and Development of Nanoparticle Characterization Methods." National Institute of Advanced Industrial Science and Technology (AIST). Available at http://www.aist-riss.jp/main/?ml_lang=en
60. Nasterlack M Zober A Oberlinner C 2008 Considerations on occupational medical surveillance in employees handling nanoparticles Int Arch Occup Environ Health 81 721 6
61. NIOSH 2009 Approaches to safe nanotechnology: managing the health and safety concerns with engineered nanomaterials Cincinnati, OH U.S. Department of Health and Human Services, Public Health Service, Centers for Disease Control and Prevention, National Institute for Occupational Safety and Health, DHHS (NIOSH). Publication No. 2009-125
62. NIOSH 2009 Current intelligence bulletin 60: Interim guidance for medical screening and hazard surveillance for workers potentially exposed to engineered nanoparticles Cincinnati, OH U.S. Department of Health and Human Services, Public Health Service, Centers for Disease Control and Prevention, National Institute for Occupational Safety and Health, DHHS (NIOSH). Publication No. 2009-116
63. NIOSH 2010 Progress toward safe nanotechnology in the workplace: a report from the NIOSH Nanotechnology Research Center Cincinnati, OH U.S. Department of Health and Human Service, Centers for Disease Control and Prevention, National Institute for Occupational Safety and Health, DHHS (NIOSH). Publication No. 2010-104
64. NIOSH 2010 Strategic Plan for NIOSH Nanotechnology Research and Guidance Filling the Knowledge Gaps Cincinnati, OH U.S. Department of Health and Human Service, Centers for Disease Control

and Prevention, National Institute for Occupational Safety and Health, DHHS (NIOSH). Publication No. 2010–105

65. NIOSH 2010 Occupational exposure to carbon nanotubes and nanofibers. Draft for public comment. Current intelligence bulletin, NIOSH Docket Number: NIOSH 161-A
66. NIOSH 2011 Current intelligence bulletin 63 Occupational exposure to titanium dioxide Cincinnati, OH U.S. Department of Health and Human Services, Public Health Service, Centers for Disease Control and Prevention, National Institute for Occupational Safety and Health, NIOSH, DHHS. Publication No. 2011-160
67. NRC 2009 Science and Decisions: Advancing Risk Assessment Committee on Improving Risk Analysis Approaches Used by the U.S. EPA, Board on Environmental Studies and Toxicology, Division on Earth and Life Studies, National Research Council of the National Academies Washington, DC The National Academies Press. ISBN-10: 0-309-12046-2
68. Oberdörster G Ferin J Lehnert BE 1994 Correlation between particle size, in vivo particle persistence, and lung injury Environ Health Perspect 102 Suppl. 5 173 9
69. Oberdörster G Sharp Z Atudorei V et al. 2004 Translocation of inhaled ultrafine particles to the brain Inhal Toxicol 16 437 45
70. Oberdörster G Maynard A Donaldson K et al. 2005 Principles for characterizing the potential human health effects from exposure to nanomaterials: elements of a screening strategy. Report of the International Life Sciences Institute Research Foundation/Risk Science Institute Nanomaterial Toxicity Screening Working Group Part Fibre Toxicol 2 8
71. Oberdörster G Oberdörster E Oberdörster J 2005 Nanotoxicology: an emerging discipline evolving from studies of ultrafine particles Environ Health Perspect 113 823 39
72. OECD 2007 Guidance on grouping of chemicals Organization for Economic Cooperation and Development, Environmental Health and Safety Publications. Series on Testing and Assessment, No. 80. ENV/JM/MONO(2007)28
73. OECD 2008 Working Party on Manufactured Nanomaterials: List of Manufactured Nanomaterials and List of Endpoints for Phase One of the OECD Testing Programme Organization for Economic Cooperation

and Development

74. OECD 2009 Report of an OECD Workshop on Exposure Assessment and Exposure Mitigation: Manufactured Nanomaterials Organization for Economic Cooperation and Development
75. OECD 2010 Report of the workshop on risk assessment of manufactured nanomaterials in a regulatory context Organisation de Coopération et de Développement Économiques
76. OECD 2010 Guidance manual for the testing of manufactured nanomaterials: OECD's sponsorship programme; first revision Organisation de Coopération et de Développement Économiques
77. Osmond-McLeod MJ Poland CA Murphy F et al. 2011 Durability and inflammogenic impact of carbon nanotubes compared with asbestos fibres Part Fibre Toxicol
78. Ostiguy C Roberge B Ménard L et al. 2009 A good practice guide for safe work with nanoparticles: The Quebec approach J Phys Conf Ser 151 012037
79. Paik SY Zalk DM Swuste P 2008 Application of a pilot control banding tool for risk level assessment and control of nanoparticle exposures Ann Occup Hyg 52 419 28
80. Pauluhn J 2010 Subchronic 13-week inhalation exposure of rats to multiwalled carbon nanotubes: Toxic effects are determined by density of agglomerate structures, not fibrillar structures Toxicol Sci 113 226 42
81. Pauluhn J 2010 Multi-walled carbon nanotubes (Baytubes®): approach for derivation of occupational exposure limit Regul Toxicol Pharmacol 57 78 89
82. Pauluhn J 2011 Poorly soluble particulates: searching for a unifying denominator of nanoparticles and fine particles for DNEL estimation Toxicology 279 176 88
83. Poland CA Duffin R Kinloch I et al. 2008 Carbon nanotubes introduced into the abdominal cavity of mice show asbestos-like pathology in a pilot study Nat Nanotechnol 3 423 28
84. Pope CA Burnett RT Thun MJ et al. 2002 Lung cancer, cardiopulmonary mortality and long term exposure to fine particulate air pollution JAMA 287 1132 41
85. Puzyn T Rasulev B Gajewicz A et al. 2011 Using nano-QSAR to predict the cytotoxicity of metal oxide nanoparticles Nat Nanotechnol

6 175 8

86. Ramachandran G Ostraat M Evans DE et al. 2011 A strategy for assessing workplace exposures to nanomaterials JOEH 8 673 85
87. Rengasamy S Eimer BC 2011 Total inward leakage of nanoparticles through filtering facepiece respirators Ann Occup Hyg 55 253 63
88. Renwick LC Donaldson K Clouter A 2001 Impairment of alveolar macrophage phagocytosis by ultrafine particles Toxicol Appl Pharmacol 172 119 27
89. Rom WN Markowitz S 2006 Environmental and Occupational Medicine 4th Lippincott Williams & Wilkins
90. Rushton EK Jiang J Leonard SS et al. 2010 Concept of assessing nanoparticle hazards considering nanoparticle dosemetric and chemical/biological response metrics J Toxicol Environ Health A 73 445 61
91. Sager TM Castranova V 2009 Surface area of particle administered versus mass in determining the pulmonary toxicity of ultrafine and fine carbon black: comparison to ultrafine titanium dioxide Part Fibre Toxicol 6 15
92. Sargent LM Shvedova AA Hubbs AF et al. 2009 Induction of aneuploidy by single-walled carbon nanotubes Environ Mol Mutagen 50 708 17
93. Sargent LM Reynolds SH Hubbs AF et al. 2011 Understanding carbon nanotube genotoxicity Toxicologist 120 A59
94. Sargent LM Hubbs AF Young SH et al. 2011 Single-walled carbon nanotube-induced mitotic disruption Mutat Res. [Epub ahead of print]
95. Schoeny RS Margosches E 1989 Evaluating comparative potencies: Developing approaches to risk assessment of chemical mixtures Toxicol Ind Health 5 825 37
96. Schubauer-Berigan MK Dahm MM Yencken MS 2011 Engineered carbonaceous nanomaterials manufacturers in the United States: workforce size, characteristics, and feasibility of epidemiologic studies J Occup Environ Med 53 Suppl. 6 S62 7
97. Schulte PA Salamanca-Buentello 2007 Ethical and scientific issues of nanotechnology in the workplace Environ Health Perspect 115 5 12
98. Schulte PA Trout DB 2011 Nanomaterials and worker health: Medical surveillance, exposure registries, and epidemiological research JOEM 53 Suppl. 6 S3 7

99. Schulte P Geraci C Zumwalde R et al. 2008 Occupational risk management of engineered nanoparticles J Occup Environ Hyg 5 239 49
100. Schulte P Geraci C Hodson L et al. 2010 Nanotechnologies and nanomaterials in the occupational setting Ital J Occup Environ Hyg 1 63 8
101. Schulte PA Kuempel ED Zumwalde R et al. 2012 Focused actions to protect carbon nanotubes workers Am J Ind
102. Semmler M Seitz J Erbe F et al. 2004 Long-term clearance kinetics of inhaled ultrafine insoluble iridium particles from the rat lung, including transient translocation into secondary organs Inhal Toxicol 16 453 459
103. Shinohara N 2011 Risk Assessment of Manufactured Nanomaterials – Fullerence (C60). Final report issued on July 2011. New Energy and Industrial Technology Development Organization (NEDO) project (P06041) "Research and Development of Nanoparticle Characterization Methods." National Institute of Advanced Industrial Science and Technology (AIST). Available at http://www.aist-riss.jp/main/?ml_lang=en
104. Shvedova AA Kisin ER Mercer R et al. 2005 Unusual inflammatory and fibrogenic pulmonary responses to single-walled carbon nanotubes in mice Am J Physiol Lung Cell Mol Physiol 289 L698 708
105. Shvedova AA Kisin E Murray AR et al. 2008 Inhalation versus aspiration of single walled carbon nanotubes in C57BL/6 mice: inflammation, fibrosis, oxidative stress and mutagenesis Am J Physiol Lung Cell Mol Physiol 295 L552 65
106. Snipes MB 1989 Long-term retention and clearance of particles inhaled by mammalian species Crit Rev Toxicol 20 175 211
107. Sobels FH 1993 Approaches to assessing genetic risks from exposure to chemicals Environ Health Perspect 101 Suppl. 3 327 32
108. Takagi A Hirose A Nishimura T et al. 2008 Induction of mesothelioma in p53+/- mouse by intraperitoneal application of multi-walled carbon nanotube J Toxicol Sci 33 105 16
109. The Royal Society, The Royal Academy of Engineering 2004 Nanoscience and nanotechnologies London The Royal Society and The Royal Academy of Engineering Available at www.nanotec.org.uk/finalReport.htm

110. Tran CL Buchanan D Cullen RT et al. 2000 Inhalation of poorly soluble particles. II. Influence of particle surface area on inflammation and clearance Inhal Toxicol 12 1113 26
111. Tsai S Huang RF Ellenbecker MJ 2010 Airborne nanoparticle exposures while using constant-flow, constant-velocity, and air-curtain isolated fume hoods Ann Occup Hyg 54 78 87
112. Wang L Mercer RR Rojanasakul Y et al. 2010 Direct fibrogenic effects of dispersed single-walled carbon nanotubes on human lung fibroblasts J Toxicol Environ Health A 73 410 22
113. Yegles M Janson X Dong HY et al. 1995 Role of fibre characteristics on cytotoxicity and induction of anaphase/telophase aberrations in rat pleural mesothelial cells in vitro. Correlations with in vivo animal findings Carcinogenesis 16 2751 8

8

Occupational Exposures to Styrene Vapor in a Manufacturing Plant for Fiber-Reinforced Composite Wind Turbine Blades

Duane Hammond, Alberto Garcia and H. AMY FENG

Division of Applied Research and Technology, National Institute for Occupational Safety and Health, Centers for Disease Control and Prevention, 4676 Columbia Parkway, Mail Stop R5, Cincinnati, OH 45226, USA

ABSTRACT

Objectives: A utility-scale wind turbine blade manufacturing plant requested assistance from the National Institute for Occupational Safety and Health (NIOSH) in controlling worker exposures to styrene at a plant that produced 37 and 42 m long fiber-reinforced wind turbine blades. The plant requested NIOSH assistance because previous air sampling conducted by the company indicated concerns about peak styrene concentrations when workers entered the confined space inside of the wind turbine blade. NIOSH researchers conducted two site visits and collected personal breathing zone and area air samples while workers performed the wind turbine blade manufacturing tasks of vacuum-assisted resin transfer molding (VARTM), gelcoating, glue wiping, and installing the safety platform.

Citation: Duane Hammond Alberto Garcia H. Amy Feng, Occupational Exposures to Styrene Vapor in a Manufacturing Plant for Fiber-Reinforced Composite Wind Turbine Blades: The Annals of Occupational Hygiene, Volume 55, Issue 6, July 2011, Pages 591–600, https://doi.org/10.1093/annhyg/mer021.

Methods: All samples were collected during the course of normal employee work activities and analyzed for styrene using NIOSH Method 1501. All sampling was task based since full-shift sampling from a prior Occupational Safety and Health Administration (OSHA) compliance inspection did not show any exposures to styrene above the OSHA permissible exposure limit. During the initial NIOSH site visit, 67 personal breathing zone and 18 area air samples were collected while workers performed tasks of VARTM, gelcoating, glue wipe, and installation of a safety platform. After the initial site visit, the company made changes to the glue wipe task that eliminated the need for workers to enter the confined space inside of the wind turbine blade. During the follow-up site visit, 12 personal breathing zone and 8 area air samples were collected from workers performing the modified glue wipe task.

Results: During the initial site visit, the geometric means of the personal breathing zone styrene air samples were 1.8 p.p.m. (n = 21) for workers performing the VARTM task, 68 p.p.m. (n = 5) for workers installing a safety platform, and 340 p.p.m. (n = 14) for workers performing the glue wipe task, where n is the number of workers sampled for a given mean result. Gelcoating workers included job categories of millers, gelcoat machine operators, and gelcoaters. Geometric mean personal breathing zone styrene air samples were 150 p.p.m. (n = 6) for millers, 87 p.p.m. (n = 2) for the gelcoat machine operators, and 66 p.p.m. (n = 19) for gelcoaters. The geometric mean of the personal breathing zone styrene air samples from the glue wipe task measured during the follow-up site visit was 31 p.p.m. (n = 12).

Conclusions: The closed molding VARTM process was very effective at controlling worker exposures to styrene. Personal breathing zone styrene air samples were reduced by an order of magnitude after changes were made to the glue wipe task. The company used chemical substitution to eliminate styrene exposure during the installation of the safety platform. Recommendations were provided to reduce styrene concentrations during gelcoating.

Keywords: alternative energy, styrene, task-based sampling, wind blade, wind turbine

INTRODUCTION

A utility-scale wind turbine blade manufacturing plant in the USA requested assistance from the National Institute for Occupational Safety and Health (NIOSH) in controlling worker exposures to styrene at a plant that produces 37 and 42 m long fiber-reinforced wind turbine blades. Prior to the first NIOSH evaluation at the wind turbine blade manufacturing plant, two separate evaluations of styrene exposures had already been conducted at the same plant by other organizations. One evaluation was an Occupational Safety and Health Administration (OSHA) compliance inspection and the other evaluation was performed internally by site personnel at the wind turbine blade manufacturing plant. The OSHA compliance inspection did not find any exposures to styrene above the OSHA permissible exposure limit (PEL) of 100 p.p.m. as an 8-h time-weighted average (TWA). Full-shift sampling conducted internally by site personnel at the wind turbine blade manufacturing plant also did not find any exposures to styrene above the OSHA PEL. However, air monitoring using MultRAE Plus direct reading instrumentation performed by site personnel during a separate evaluation indicated that peak styrene concentrations were above short-term exposure limits (STELs) during several tasks. Following the OSHA and company evaluations, NIOSH researchers conducted a walk-through evaluation and observed several tasks with high potential for short-term styrene exposure while also noting that the same workers also spent a considerable amount of time working with dry materials with minimal exposure to styrene. Based on this information, NIOSH researchers decided to conduct task-based sampling and look for opportunities to recommend engineering controls to reduce concentrations of styrene in the air for four tasks.

During the initial in-depth site visit, NIOSH researchers collected 67 personal breathing zone and 18 area air samples while workers performed the wind turbine blade manufacturing tasks of vacuum-assisted resin transfer molding (VARTM), gelcoating, glue wiping, and installing the safety platform. After the initial site visit, the company made changes to the design of the molds that eliminated the need for workers to enter the wind turbine blade to wipe off excess glue. NIOSH researchers conducted a follow-up site visit 3 months later and collected 12 personal breathing zone and 8 area air samples from workers performing the modified glue wipe task.

Styrene is a fugitive emission when it evaporates from resins, gelcoats, solvents, and surface coatings commonly used in the manufacturing process for fiber-reinforced plastics (FRP) and can present an inhalation hazard to workers handling these materials. The polyester resins used at the studied wind turbine blade manufacturing plant contained between 36 and 42% styrene.

The following exposure criteria include both full-shift and short-term criteria even though all sampling during the present evaluation was task based. The NIOSH recommended exposure limit for styrene is 50 p.p.m. as a 10-h TWA, with a 15-min STEL of 100 p.p.m. (NIOSH, 2004). These recommendations are based upon reported central nervous system effects, eye irritation, and respiratory irritation effects. The NIOSH immediately dangerous to life or health (IDLH) value for styrene of 700 p.p.m. is based on acute inhalation toxicity in humans. The OSHA PEL for styrene is 100 p.p.m. as a TWA. The OSHA ceiling limit for styrene is 200 p.p.m. with a 600 p.p.m. 5-min maximum peak in any 3 h. The American Conference of Governmental Industrial Hygienists (ACGIH®) revised its threshold limit value (TLV®) in 1997 and recommends styrene be controlled to 20 p.p.m. TWA with a 40 p.p.m. STEL (ACGIH®, 2009). The TLV® is based on a number of health effects of low styrene exposure such as ototoxicity, central and peripheral neurologic, optic, and irritant actions in humans (ACGIH®, 2001). The Swedish Work Environment Authority (SWEA) has an occupational exposure level limit value for styrene of 20 p.p.m. and a short-term value of 50 p.p.m. (SWEA, 2005).

In February 1996, the Styrene Information and Research Center (SIRC) and three other styrene industry trade associations—American Composites Manufacturers Association, National Marine Manufacturers Association, and the International Cast Polymer Association—entered into an arrangement with OSHA to voluntarily adhere to the 50-p.p.m. level set by the 1989 update of the OSHA PEL that was later vacated by court order (OSHA, 1989). OSHA announced the voluntary agreement in a 1996 newsletter (OSHA, 1996). The SIRC encouraged its members to continue to comply with the 50-p.p.m. standard as an appropriate exposure level for styrene, regardless of its regulatory status (SIRC, 1996).

Facility Description

At the time of the evaluation, the facility was operating multiple 10- and 12-h overlapping shifts to manufacture wind turbine blades 24 h day^{-1}, 365 days per year. Workers performing the glue wipe task were working 12-h shifts and workers performing gelcoating, VARTM, and installation of the safety platform worked 10-h shifts. Approximately 600 of the plant's

940 employees worked in areas where there was potential for exposure to styrene vapor. At the time of the evaluation, approximately eight wind turbine blades were produced per day using 16 mold tools with a cycle time near 24 h. The manufacturing operations took place in two buildings on the ~50 000 m^2 (12.4 acre) property. Each building contained ~9000 m^2(96 900 ft^2) of manufacturing floor space. The VARTM, gelcoating, and glue wiping tasks were performed in Building 1 and installation of the safety platform was performed in Building 2.

The supply air flow rates from the four air handling units in each of the buildings were provided by facility representatives and were ~87 $m^3\ s^{-1}$ in Building 1 and ~94 $m^3\ s^{-1}$ in Building 2. The plant used direct reading MultiRAE Plus monitors near each task to measure for potential buildup of styrene vapor. The MultiRAE Plus monitors were calibrated to styrene and generally located between the process and the building general exhaust ventilation.

The dilution ventilation supply system for the manufacturing space in both buildings consisted of fabric sock air distribution systems located near the ceiling. Exhaust locations for each dilution ventilation system were generally located along the walls in both buildings. Additional exhaust vents were located in the floor in Building 1. The exhaust vents in the floor were originally located to be at the ends of the wind turbine blades; however, as product demands required longer wind turbine blades, the ends of the blades extend beyond the location of the vents. According to plant management, the supply air flow rate for each system was greater than that of the exhaust air to keep the plant under positive pressure. The supply air system delivered 100% outside air, heated or cooled, as needed, so there was no recirculation.

Process Description

FRP wind turbine blades at this plant were manufactured using a closed molding technology referred to as VARTM. VARTM is a form of resin transfer molding that uses vacuum to offset some of the injection pressure. Compared to open molding, closed molding technology should significantly reduce environmental emissions and worker exposure to styrene. However, the gelcoating portion of most closed molding operations is still performed in an open mold and represents a potential source of exposure (Hammond *et al.*, 2007).

FRP wind turbine blades were manufactured at this plant starting from the outside of the blade to the center. Molds were prepared with the application of a release agent that allowed the part to separate from the mold when it was finished. After the release agent was applied, the part was ready for gelcoating. The gelcoat was a pigmented polyester resin that contained styrene. The gelcoat was sprayed on the mold where it hardened to produce a smooth outer surface. The sample duration for the gelcoating task was ~45 min and the equipment consisted of a hand spray tool with hoses connected to a gelcoating machine on wheels that was pushed along manually. Each half mold tool required three workers for the spray application of gelcoat. The gelcoater walked backwards along the concave side of the mold while spraying the gelcoat. Two other workers walked along the ground next to the sprayer to operate the gelcoat machine. Workers performing the gelcoating tasks at the evaluated plant wore powered air purifying respirators with organic vapor cartridges. After gelcoating, the workers took a break and waited for the concentration of styrene in the air to drop <20 p.p.m. as measured by direct reading monitors. After the concentrations dropped <20 p.p.m., the same workers that applied the gelcoat removed their respirators and began preparing core materials next to the mold. After gelcoating, the lamination task began with the placement of the glass fibers and core material. The mold was then covered and sealed with a vacuum film. Vacuum lines were attached to the mold to assist the flow of resin to saturate and bond the fiber and core material. Workers performing the VARTM task monitored various flow, temperature, and other parameters and did not come in contact with the resin. The vacuum film over the mold also prevented the release of styrene vapor. The sample duration for the VARTM task was ~2 h. This closed molding VARTM method was used instead of traditional open molding methods of manufacturing fiberglass parts. The VARTM process is shown in Fig. 1.

Using the VARTM method, the wind turbine blades were laminated in two shells before the sandwich web was installed. The sandwich web ran almost the entire length of the wind turbine blade and provided structural reinforcement. Glue containing 34% styrene by weight was applied to the edges of the shell and web, and one shell was lifted and fixed to the other half to assemble the wind turbine blade. After the two shells were pressed together, workers entered the confined space inside of the wind turbine blade to wipe any excess glue that was pressed out.

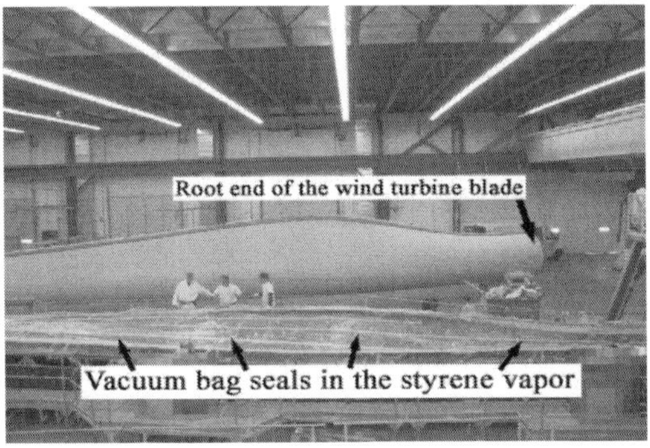

Figure 1: The vacuum film over the mold for the VARTM process.

Workers crawled around inside of the three sections of the wind turbine blade and wiped excess glue using a plastic tool and placed the glue into a bucket. The excess glue was located mostly in areas where the two half shells were placed together and along the edges next to the surface where the sandwich web contacted the shell. The workers were wearing sampling equipment [and respirators] to measure personal breathing zone concentrations of styrene. The glue wipe task took 15 min but the sampling duration was ~40 min to allow time for the workers to complete tasks outside of the wind turbine blade before and after wiping glue. Figure 2 shows workers inside of the wind turbine blade wiping the glue with a plastic tool and putting the glue into a bucket.

Figure 2: Workers wiping glue inside of the wind turbine blade.

After the initial NIOSH site visit, the company made changes to the mold design that eliminated the need for workers to enter the wind turbine blade during the glue wipe task. Design changes tightened dimensional tolerances of the mold so that less glue was pressed out when the two half wind turbine blades were joined together. Another change included a fiberglass flange on the inside of the wind turbine blade to catch glue that was pressed out and hold it against the surface where the two parts joined. The fiberglass flange is shown in Fig. 3. These changes eliminated the need for workers to enter the wind turbine blade during the glue wipe task. Instead, workers stood outside of the wind turbine blade and used a long handled tool to wipe the excess glue near the open end of the blade.

Figure 3: Fiberglass flange to catch glue after the two half wind turbine blades are joined together.

After the two shells or half wind turbine blades were joined together, the wind turbine blade was removed from the mold for cutting, grinding, and sanding of the outside edge to provide a smooth finish. After all grinding and sanding tasks were completed, a safety platform was installed near the root of the wind turbine blade. The safety platform was installed using styrene resin and glassed in by hand using plastic tools and a bucket of resin. The platform was installed as two half disks with some workers applying resin from the inside of the blade while other workers applied resin from the outside. The sample duration for the installation of the safety platform was ~1 h. An installed safety platform is shown in Fig. 4.

Figure 4: Safety Platform glassed in as a disk shaped fiberglass part near the root of the wind turbine blade.

METHODS

Air Sampling for Styrene

All personal breathing zone and area air styrene samples were collected for the duration of each individual task where styrene was used. Since workers performed some tasks with styrene and some without styrene during their shift, the decision was made to conduct task-based personal breathing zone and area sampling instead of full-shift sampling. Also, air sampling data from a prior OSHA compliance inspection did not find any exposures above the OSHA PEL of 100 p.p.m. for styrene. Additional air sampling data collected by the company raised concerns about peak exposures during several tasks.

Personal breathing zone and general area air samples for styrene were collected and analyzed in accordance with NIOSH Method 1501 (NIOSH, 1994). Samples were collected on SKC sorbent tubes (Model number 226-01, Anasorb CSC, Coconut Charcoal, Lot #2000). After breaking the sealed ends, each tube was connected to a low flow SKC Pocket Pump (SKC Inc.) set at a flow rate depending on the sample time of the task. The nominal flow rate was set to 100 ml min^{-1} during sampling of tasks for the gelcoaters, millers, gelcoat machine operators, and workers performing the glue wiping tasks. The nominal flow rate was set to 60 ml min^{-1} for the tasks of VARTM and the installation of the safety platform. The pumps' actual flow rates were calibrated before and after sampling using a DryCal® primary calibration standard (BIOS, Butler, NJ, USA). For personal breathing zone air samples, the air inlet of the sampling apparatus was secured in each worker's breathing zone with a lapel clip, and the battery-powered pump clipped to the worker's belt. In addition, field blank samples were collected to ensure that the sample media was not contaminated and to account for any variance in sample preparation. The analyses of the charcoal tube samples for styrene were performed by Bureau Veritas North America, Inc., in Novi, Michigan. The limit of detection and limit of quantification for styrene for this sample set was 8 μg per sample and 180 μg per sample, respectively.

General area air samples were collected using the same type of sampling apparatus and flow rates as used for the personal air sampling. These samples were placed in stationary locations near each evaluated task.

Once the sample results were received from the analytical laboratory, the styrene breathing zone concentrations and general area concentrations were calculated using equation (1).

$$C = \frac{m}{V \times 4.26}, \qquad (1)$$

Where C = styrene concentration, parts per million; m = mass of styrene per sample, micrograms; V = volume of air sample, liter; and 4.26 = conversion constant micrograms per liter to parts per million (NIOSH, 1994)

RESULTS

The distribution of all samples was checked for normality using the normal plot, frequency histogram, and Shapiro–Wilk's test and the results indicated that the data were log-normally distributed. Subsequently, all data were log transformed for statistical analysis. Geometric mean, geometric standard deviation, 95% confidence limits, and sample size for comparison of personal breathing zone and general area air styrene samples are included in Table 1 for the initial and follow-up site visits.

Table 1: Personal and area sample statistical results for styrene vapor

Job description	Sample type	Geometric mean (p.p.m.)	Geometric standard deviation	Geometric upper 95% confidence limit	Geometric lower 95% confidence limit	*n*
Infuser (VARTM)	Personal	1.8	1.5	2.2	1.5	21
Infusion (VARTM)	Area	1.5	1.1	1.9	1.2	3
Safety platform	Personal	68	4.3	410	11	5
Safety platform	Area	4.2	1.5	11	1.6	3
Gelcoat machine	Personal	87	2.0	a	a	2
Gelcoat	Personal	66	2.6	100	42	19
Miller	Personal	150	1.4	220	100	6
Gelcoat	Area	47	2.4	91	24	9
Glue wipe (first site visit)	Personal	340	2.0	510	230	14
Glue wipe (second site visit)	Personal	31	3.2	65	14	12
Glue wipe area outside of the blade (first site visit)	Area	18	1.3	37	8.6	3
Glue wipe area outside of the blade (second site visit)	Area	11	1.6	17	9.1	8

[a]No meaningful confidence limit due to small sample size and high variation

The personal breathing zone samples for styrene from workers performing the glue wipe task inside the wind turbine blade were higher than any other evaluated task with a geometric mean concentration of 340 p.p.m. The geometric mean air styrene concentration from area samples collected outside of the wind turbine blade during the glue wipe task was 18 p.p.m. The geometric mean of personal breathing zone task-based air sampling from the glue wipe task collected during the follow-up evaluation after the design change was 31 p.p.m. This is approximately an order of magnitude lower than the task-based sampling results from the glue wipe task collected during the initial evaluation. This reduction in personal breathing zone styrene concentrations was a result of the design change that eliminated the need for workers to enter the space inside of the wind turbine blade during the glue wipe task. The workers that were sampled for the follow-up survey included the same workers that would have otherwise been working inside of the wind turbine blade. Instead, these workers were performing some glue wiping while standing outside of the wind turbine blade and reaching inside of the blade with a long handled tool.

The lowest personal breathing zone sampling results for styrene were measured among the 21 workers performing the VARTM task. All personal breathing zone and general area styrene air samples measured during the VARTM task were <5 p.p.m. for the ~2-h task time sampled.

The geometric mean of the personal breathing zone task-based air sampling for styrene from the gelcoaters, gelcoat machine operators, and millers were 66, 87, and 150 p.p.m., respectively, for the ~45-min task time sampled. The highest two individual personal breathing zone styrene air samples collected from gelcoating workers were 250 p.p.m. for a 42-min sampling time and 230 p.p.m. for a 45-min sampling time. Actual worker exposures were likely lower than measured concentrations since at the time of the survey, all gelcoaters, millers, and gelcoat machine operators wore either half-mask or powered air purifying respirators with organic vapor cartridges that were changed out daily. Under the 'NIOSH Respirator Decision Logic', any air purifying half-mask respirator equipped with appropriate gas/vapor cartridges has an assigned protection factor of 10 (NIOSH, 1987). Any powered air purifying respirator with a loose-fitting hood or helmet equipped with appropriate gas/vapor cartridges has an assigned protection factor of 25.

Five personal breathing zone samples for styrene were measured from workers installing the safety platform. Three of the five air samples were measured from workers who were standing inside of the wind turbine blade while applying resin to the safety platform and two from workers outside the wind turbine blade while applying resin to the other side of the safety platform. For the 1-h task time sampled, the three personal breathing zone samples measured from workers standing inside the wind turbine blade from the safety platform were 92, 156, and 316 p.p.m., while personal air styrene samples measured from workers who were outside the wind turbine blade from the safety platform were 46 and 7 p.p.m. Actual exposures were much lower since workers wore full-face piece respirators with organic vapor cartridges that were changed out daily. Any air purifying full-face piece respirator equipped with appropriate gas/vapor cartridges has an assigned protection factor of 50.

The personal breathing zone results were highly variable for air samples collected from workers performing tasks near the opening to the wind turbine blade for the glue wipe and safety platform tasks. The gradient of styrene concentrations changed more than an order of magnitude in the short distance between the low concentrations outside of the wind turbine blade and the much higher concentrations inside of the wind turbine blade. Therefore, it is expected that workers performing tasks at locations within this gradient might have highly variable personal breathing zone results. It is possible that the distance between the breathing zone and resin, proximity to the opening of the wind turbine blade, or other factors contributed to the variability. There is some evidence in the public literature to suggest that high variability in styrene air sampling data can be expected. A study of styrene exposures at a boat manufacturing plant found significant differences in styrene exposures when four monitors were placed at different locations around the breathing zone of each worker (Malek et al., 1999).

DISCUSSION

The task-based sampling performed in this study provided information on which tasks could benefit most from introducing controls. But even sampling the glue wipe task presented some difficulties in attempting to measure concentrations only during the task time. The glue wipe task inside of the wind turbine blade took 15 min. However, it was not possible to start and stop samples immediately when workers entered and exited the wind turbine blades due to the need for workers to put on or take off personal protective equipment or perform additional tasks outside the wind turbine blade while sample pumps were still operating. The results for the glue wipe personal breathing zone samples collected during the initial site visit as presented in Table 1 include the 15 min each worker spent inside the wind turbine blade along with the time spent outside the wind turbine blade during the glue wipe task. For discussion purposes, it is possible to calculate an adjusted value for each of the 14 glue wipe personal breathing zone air samples that subtracts the time each worker spent outside of the wind turbine blade during the glue wipe task. This calculation of an adjusted value is being made in this particular case to estimate the 15-min breathing zone concentration using the available sample result as if it had been possible to obtain the 15-min sample inside the confined space of the wind turbine blade. The adjusted values are calculated by assuming worker exposures during the glue wipe task were 18 p.p.m. for the portion of the sample time that exceeded 15 min. The value of 18 p.p.m. is used since the three area samples collected outside of the wind turbine blade had a geometric mean of 18 p.p.m. The adjusted values were calculated using equation (2):

$$C_i = \frac{C_t \times t_t - C_o \times t_o}{t_i}, \tag{2}$$

Where C_i = styrene concentration when the worker was inside of the wind turbine blade, parts per million; C_t = styrene concentration for the task time sampled, parts per million; C_o = styrene concentration for the time the worker spent outside of the wind turbine blade, parts per million; t_i = time the worker spent inside of the wind turbine blade, minutes; t_t = task time sampled, minutes; t_o = time the worker spent outside of the wind turbine blade, minutes.

For example, the first glue wipe personal breathing zone styrene air sample collected from the glue wipe task was 603 p.p.m. for a 29-min sample. The following example calculation estimates the concentration during the 15 min the worker spent inside the wind turbine blade by subtracting out the 14 min spent outside of the wind turbine blade at 18 p.p.m.

$$C_i = \frac{(603\,\text{p.p.m.} \times 29\text{min}) - (18\,\text{p.p.m} \times 14\text{min})}{15\text{min}}$$
$$= 1149\,\text{ppm}. \tag{3}$$

Based on this sample calculation, is it possible to estimate that the concentration for the first personal breathing zone air sample for styrene was 1149 p.p.m. (0.11%) for the 15 min that the worker spent inside of the wind turbine blade. After applying this calculation to each of the 14 styrene personal breathing zone air samples collected from the glue wipe task during the initial site visit, the geometric mean is calculated at 970 p.p.m. for the 15 min that the workers spent inside the wind turbine blade during the glue wipe task. Glue wipe personal breathing zone samples for styrene before adjustment indicated several samples were either approaching or higher than the established NIOSH IDLH exposure level for styrene of 700 p.p.m. The adjusted personal breathing zone calculations indicate that 12 of the 14 values were >700 p.p.m.

OSHA Compliance Assistance Guidelines for Confined and Enclosed Spaces and Other Dangerous Atmospheres [29 CFR 1915.12(b)(3)] considers that atmospheres with a 'concentration of flammable vapors ≥10% of the LEL are considered hazardous when located in confined spaces'. The LEL for styrene is 0.9% or 9000 p.p.m. Concentrations should be <900 p.p.m. in order to remain <10% of the LEL. When the data from the glue wipe task were adjusted for the time workers were inside of the wind turbine blade, 11 of the 14 adjusted values were >900 p.p.m.

NIOSH Respirator Decision Logic for IDLH Conditions

The company had a respiratory protection program. However, during the initial site visit, workers performing glue wipe tasks wore powered air purifying respirators which is not consistent with NIOSH respirator decision logic for IDLH atmospheres. The current NIOSH definition for an IDLH condition, as given in the NIOSH Respirator Decision Logic, is 'an exposure condition that poses a threat of exposure to airborne contaminants when that exposure is likely to cause death or immediate or delayed permanent adverse health effects or prevent escape from such an environment'. NIOSH has established an IDLH exposure level for styrene of 700 p.p.m. The IDLH is considered a maximum level above which only a highly reliable breathing apparatus providing maximum worker protection is permitted. Any appropriate approved respirator may be used to its maximum use concentration up to the IDLH concentration. Under the NIOSH Respirator Decision Logic, only 'highly reliable' respirators (i.e. the most protective respirators) would be selected for IDLH conditions (NIOSH, 1987). These highly reliable respirators include a pressure demand, full-face piece self-contained breathing apparatus (SCBA) or a pressure demand, full-face piece supplied air respirator (SAR) in combination with an auxiliary pressure demand, and full-face piece SCBA. The auxiliary SCBA must be of sufficient duration to permit escape to safety if the air supply is interrupted. An auxiliary unit means that the SAR unit includes a separate air bottle to provide a reserve source of air should the airline become damaged. The auxiliary unit shares the same mask and regulator and enables the SAR to function as an SCBA if needed. During the follow-up evaluation of the glue wipe task, workers wore powered air purifying respirators with organic vapor cartridges which was consistent with the NIOSH respirator decision logic since they were no longer entering the confined space or IDLH atmosphere due to the design change.

CONCLUSIONS

Initial Site Visit

The closed molding VARTM task was very effective at controlling worker exposures to styrene as demonstrated by the air samples collected during the initial site visit. Air sampling results from the gelcoating, glue wipe, and safety platform installation tasks indicated that control measures were needed to reduce styrene concentrations.

During the initial site visit, the plant was using respiratory protection combined with an administrative control to reduce average styrene exposures of workers during the gelcoating task. The administrative control required workers to leave the room and take a break after gelcoating was finished and return when area concentrations for styrene decreased <20 p.p.m. Recommendations were made that the administrative control effort should continue and all workers in the gelcoating area should be required to leave the room to take advantage of the administrative control. Recommendations were also made to continue the use of organic vapor charcoal filter respirators during gelcoating. Recommendations were made to consider installing traveling or moving air systems to provide local exhaust ventilation during gelcoating. The moving air systems could be designed to supply fresh air near the breathing zone of the worker and exhaust air downstream of the source with air flows low enough not to interfere with the application of gelcoat. Worker exposures to styrene could be reduced further by implementing robotic gelcoating. In-mold gelcoating technology may also be applicable to VARTM and large components (Summerscales *et al.*, 2010).

Follow-up Site Visit

At the time of the follow-up evaluation, the design changes to the molds eliminated the need for workers to enter the wind turbine blade during the glue wipe task. This change was very important since it dramatically reduced exposures to styrene for workers that were previously entering the wind turbine blade. Although the changes to the glue wipe task dramatically reduced worker exposures to styrene, some workers were observed standing outside of the wind turbine blade while leaning in with a pole to wipe glue near the opening. Other workers standing near the opening to the wind turbine blade for training or other purposes did not consistently wear respiratory protection. The pole used by the worker was flimsy which increased the likelihood that the worker would lean in and cross the plane of the confined space while wiping glue. Recommendations were made to the company that better tools should be provided to the glue wipe workers along with training to keep their breathing zone out of the confined space during the task. Recommendations were also made to continue using organic vapor charcoal filter respirators for workers performing the glue wipe task or standing near the opening to the wind turbine blade.

After the follow-up site visit, management along with plant engineers began investigating pneumatic conveying air systems that follow the gelcoater along the wind turbine blade. The traveling or moving air system could possibly be tied into existing conveying systems over each mold.

Since the time of the evaluation, the company initiated a chemical substitution to eliminate styrene exposures during the installation of the safety platform. A product called Sika Flex replaced the styrene resin used to secure the safety platform inside of the wind turbine blade. Although the new adhesive did not contain styrene, the new Sika Flex product contained xylene. Recommendations were provided that sampling for xylene vapor should be conducted to make sure that it does not replace one hazard with another and present an exposure hazard for workers. The OSHA PEL for xylene is 100 p.p.m. The NIOSH recommended STEL for xylene is 150 p.p.m. Recommendations were made to consider the STEL in addition to the PEL when sampling for xylene.

The task-based sampling approach was effective in identifying tasks that could benefit from the introduction of engineering controls to reduce styrene concentrations during tasks such as glue wipe, gelcoating and the installation of the safety platform. The modifications to the glue wipe process dramatically reduced personal breathing zone concentrations of styrene vapor for the task time sampled. Chemical substitution was used to eliminate styrene exposure during the installation of the safety platform. Recommendations were provided to reduce styrene concentrations during the gelcoating process. The VARTM technology effectively reduced personal breathing zone styrene concentrations. Other companies who manufacture FRP parts should consider closed molding technologies such as VARTM to reduce occupational exposures.

REFERENCES

1. ACGIH®, Documentation of threshold limit values and biological exposure indices: TLV® for Styrene , 2001Cincinnati, OHACGIH® Available at http://www.acgih.org/store/BrowseProducts.cfm?type=cat&id=16. Accessed 4 April 2011
2. ACGIH®, 2009 TLVs® and BEI s®: threshold limit values for chemical substances and physical agents & biological exposure indices , 2009Cincinnati, OHACGIH® Available at http://www.acgih.org/tlv/. Accessed 2 August 2010
3. Hammond DR, Carlo RV, Garcia A, et al. , In-depth survey report: a re-evaluation of styrene and noise in the fiberglass reinforced plastic boat manufacturing industry at Island Packet Yacht , 2007 NIOSH Report No. EPHB 306-18a. Cincinnati, OH. Available at www.cdc.gov/niosh/surveyreports/pdfs/ECTB-306-18a.pdf. Accessed 2 August 2010
4. Malek RF, Daisy JM, Cohen BS. Breathing zone concentration variations in the reinforced plastic industry; field measurements in a boat manufacturing plant, Appl Occup Environ Hyg , 1999, vol. 14 (pg. 777-84)
5. NIOSH, NIOSH respirator decision logic , 1987Cincinnati, OHNIOSH Publication No. 87-108. Available at http://www.cdc.gov/niosh/docs/2005-100/pdfs/05-100.pdf. Accessed 4 April 2011
6. Eller PM, Cassinelli ME. NIOSH. (1994)Hydrocarbons, aromatic: method 1501 (supplement issued 5/15/85), NIOSH manual of analytical methods 4thCincinnati, OHNIOSH Publication No. 94-113. Available at http://www.cdc.gov/niosh/nmam/. Accessed 2 August 2010
7. NIOSH, NIOSH pocket guide to chemical hazards and other databases-REL for styrene , 2004Cincinnati, OHNIOSH Publication No. 2004-103. Available at http://www.cdc.gov/niosh/npg/. Accessed 2 August 2010
8. OSHAU.S. Department of Labor, "air contaminants: final rule," Federal Register 54, 2332–2983, 1989 Available at http://www.osha.gov/pls/oshaweb/owadisp.show_document?p_table=FEDERAL_REGISTER&p_id=12908. Accessed 4 April 2011
9. OSHAOSHA national news release. U.S. Department of Labor Office of Public Affairs: News Release USDL, 96-77. Released March 1, 1996, 1996 Available at http://www.osha.gov/pls/oshaweb/owadisp.show_document?p_table=NEWS_RELEASES&p_id=377. Accessed 4 April 2011

10. SIRCStyrene Information and Research Center. Letter from the Assistant Secretary of the U.S. Department of Labor, 1996 Available at http://www.acmanet.org/ga/osha_styrene_agreement_docs_1996.pdf. Accessed 2 August 2010
11. Summerscales J, Hoppins C, Anstice P, et al. , In-mould gel coating for resin transfer moulding, The 10th International Conference on Flow Processes in Composite Materials (FPCM10) , 2010Monte Verità, Ascona, CH
12. SWEA, Statute Book of the Swedish Work Environment Authority: occupational exposure limit values and measures against air contaminants. (AFS 2005:17) , 2005Solna, SwedenSwedish Work Environment Authority Available at http://www.av.se/dokument/inenglish/legislations/eng0517.pdf. Accessed 2 August 2010

9

Risks, Release and Concentrations of Engineered Nanomaterial in the Environment

Bernd Giese[1,2], Fred Klaessig[3,4], Barry Park[5], Ralf Kaegi[6], Michael Steinfeldt[1], Henning Wigger[1,7], Arnim von Gleich[1] , and Fadri Gottschalk[8]

[1]University of Bremen, Faculty of Production Engineering, Department of Technology Design and Technology Development, Badgasteiner Str, 1 28359, Bremen, Germany
[2] University of Natural Resources and Life Sciences, Institute of Safety and Risk Sciences, Borkowskigasse 4, 1190, Vienna, Austria
[3] Pennsylvania Bio Nano Systems, Doylestown, Pennsylvania, 18901, United States
[4]Center for Environmental Implications of Nanotechnology (UC CEIN), University of California Santa Barbara, Santa Barbara, California, 93106-5131, United States
[5]GBP Consulting Ltd, Purton, UK

Citation: Bernd Giese, Fred Klaessig, Barry Park, Ralf Kaegi, Michael Steinfeldt, Henning Wigger, Arnim von Gleich & Fadri Gottschalk, 2018, Risks, Release and Concentrations of Engineered Nanomaterial in the Environment, Open Access Int. Scientific Reports 8, Article number: 1565 (2018), https://www.nature.com/articles/s41598-018-19275-4.

[6] Eawag, Swiss Federal Institute of Aquatic Science and Technology, Überlandstrasse 133, 8600, Dübendorf, Switzerland
[7]Empa, Swiss Federal Laboratories for Materials Science and Technology, Lerchenfeldstrasse 5, CH-9014, St. Gallen, Switzerland
[8] ETSS AG, Engineering, technical and scientifc services, CH7558, Strada, Switzerland

ABSTRACT

For frequently used engineered nanomaterials (ENMs) CeO_2-, SiO_2-, and Ag, past, current, and future use and environmental release are investigated. Considering an extended period (1950 to 2050), we assess ENMs released through commercial activity as well as found in natural and technical settings. Temporal dynamics, including shifts in release due to ENM product application, stock (delayed use), and subsequent end-of-life product treatment were taken into account. We distinguish predicted concentrations originating in ENM use phase and those originating from end-of-life release. Furthermore, we compare Ag- and CeO_2-ENM predictions with existing measurements. The correlations and limitations of the model, and the analytic validity of our approach are discussed in the context of massive use of assumptive model data and high uncertainty on the colloidal material captured by the measurements. Predictions for freshwater CeO_2-ENMs range from 1 pg/l (2017) to a few hundred ng/l (2050). Relative to CeO_2, the SiO_2-ENMs estimates are approximately 1,000 times higher, and those for Ag-ENMs 10 times lower. For most environmental compartments, ENM pose relatively low risk; however, organisms residing near ENM 'point sources' (e.g., production plant outfalls and waste treatment plants), which are not considered in the present work, may be at increased risk.

INTRODUCTION

Owing to unique qualities and manifold variation possibilities of engineered nanomaterials (ENMs), numerous products and processes involving ENMs have been developed during the past decades1,2,3,4. Medium–term forecasts predict a constant growth of ENM production5. Without much doubt, the volume and variety of ENMs released into the environment during manufacture, transport, use, and disposal will increase accordingly6. This trend demands a comprehensive analysis of present and future toxin exposure in order to prepare for potential preventive measures1. Recent indications on the toxicity of already established ENMs reveal that widespread use of a substance should not be mistaken for evidence that the substance does no harm[7].

As producers engage in abrasion, burning, cleaning, or degradation of matrices containing ENMs, all environmental compartments are likely to be exposed to ENMs. To date, the investigation of toxicological and, in particular, ecotoxicological effects of ENMs has lagged behind the study of their technical qualities and the development of new applications[8]. Synthetic amorphous silicas (SiO_2-ENMs) have been used since the middle of the last century and are produced in large amounts[9]. In recent years, leading producers of SiO_2-ENMs have increased their production capacities, in particular for mass applications such as tires[10,11]. Nanosilver (Ag-ENM) has been used for medical purposes since the beginning of the last century[12]. Similar to SiO_2-ENMs, Ag-ENM is quite versatile in its applications. It has antimicrobial properties and is a component of many consumer products[13]. Unlike SiO_2-ENMs, however, Ag-ENM has been studied broadly[14,15]. Turning to CeO_2-ENM, a large percentage is used for chemical-mechanical planarization (CMP), and CeO_2-ENM has also been used in automotive catalytic converters (ACCs) and as a fuel additive[16,17]. Given such established and diffuse applications, an analysis of the environmental distribution and concentration of the applied nanoparticles is advisable.

Compared with the toxicological examination of ENMs, few data are available about their actual release[18,19,20]. In terms of release, Ag-ENM is again the most studied ENM so far[21,22,23,24]. From a regulatory point of view, data on ENM environmental release and exposure are required to estimate the associated risk. Regardless of whether new or established ENMs are examined, models of exposure and their combination with toxicological data can contribute to a prospective risk assessment according to the precautionary principle[25,26,27]. However, this only applies to the extent that the uncertainty of exposure and toxicology models can be managed and their results validated based on reliable model input data; such data are mostly missing in ENM contexts[18]. Our stochastic approach takes into account, when necessary, relatively large uncertainty ranges from all model input parameters in the form of probability distributions, thereby generating the largest possible model output spectrum. Such uncertainty computation includes virtually all environmental release parameters as well as all raw toxic data used, effectively ensuring a conservative risk calculation.

The preliminary exposure analysis used in the present study combines elements of a scientific model of the fate and dispersal of ENMs with elements of an economic model of ENM use. Both models have separate sources of uncertainty. In principle, empirical measurements are subject to uncertainties associated with the precision, accuracy, and test relevancy of a scientific model. In particular, the analytical limitations of measurements in discriminating between natural and engineered colloids are potentially insurmountable[23]. Hence, to date, the results of models have been difficult to validate. The economic model is a snapshot in time of a dynamic marketplace that responds to price, availability, regulations, labeling, customer preferences, and other factors related to competitive materials. However, there is considerable uncertainty in the economic model, resulting from antitrust, patents, proprietary interests, and competitive advantage. Material flow analysis substitutes release data for the revenue of a market model[28]. These models often rely on assumptions that are known to be incorrect, such as balanced budgets or efficient markets. In such cases, the known incorrect assumptions serve to highlight or isolate parameters that would otherwise be lost in economic "noise"[29,30]. For the purpose of this study, which examines the individual's potential for nanomaterial exposure in the German economy, two arbitrary yet reasonable decisions were necessary. The first decision was to limit the discussion to manufactured materials, thereby excluding the consideration of incidental and natural nanomaterial exposures such as pollution. The second decision involved the selection of a mass transfer model. Although reviewed literature[31] provided justifications for each decision, it was necessary to retain simplifying assumptions with regard to the broad question of the potential for adverse health effects arising from these calculated exposures. In particular, care should be taken in comparing risk quantifications based on non-validated calculations[18] with highly controversial minimal or exposure levels that assume no effects[32].

Against this critical epistemic background, all past and future investigations of release and exposure in the field of ENMs confront a number of fundamental practical challenges. Basic obstacles include missing data on production volumes of ENMs and on the share used in different products. Meanwhile, some well-known databases provide information about products containing ENMs, but, with the exception of the Nanowerk Database[33], they focus on consumer products such as household articles, cosmetics, or textiles. Thus, important, mass-relevant products like tires are probably underrepresented.

Because market studies are expensive to procure, access to them is limited. Moreover, even if the producers provided accurate datasets to market research organizations, such studies can provide only the quantities being produced at a given time. Furthermore, different studies can arrive at very different sets of results, which suggests that producers "guesstimate" amounts being produced and used (cp. suppl. inform. on global production volumes in Holden, *et al.*)28. Hence, these organizations hide their real data, which, unfortunately, makes conducting research even more difficult. Existing studies on ENMs are no less accurate than are those on other material segments. Because ENMs are relatively new, little history is available on which to base effective judgments about present and future production quantities (cp. Table S1b and market projections in Sun, *et al.*)34.

Given the forecasted increasing use of CeO_2-, SiO_2- and Ag-ENMs in a variety of applications5, the goals for the present work were to model time-dependent past, current, and future use, release, and dispersal in natural and technical environments. Information on the annual global production volume of CeO_2-, SiO_2- und Ag-ENMs, as well as their annual trends, was researched based on a standardized questionnaire survey that involved producers, dealers, and research institutions in the field of nanomaterials. These investigations of use volumes were complemented by collecting and evaluating use data from available literature sources. Relevant product categories and the respective fractions of ENMs used for documented applications have been determined by a literature survey and an assessment of databases for nanomaterial applications. Product-life time-based transfer coefficients and factors describing the conversion of ENM into compounds without a nano-specific character enabled a time-dependent analysis of ENM transport and environmental release. The dynamic model used in this work takes into account the delay due to product storage and multi-year life cycles, as well as differences in transport and stability of ENMs. It allows an estimation of present and future mass-based release flows, dispersion into nature and the technosphere, and average environmental concentrations. The results may be used in future studies for covering spatial concentration heterogeneities by locally distributing the total German release. To enable an estimation of potential risks due to hazardous effects in soil and surface waters, a preliminary risk assessment was conducted by analyzing the overlap of predicted environmental concentration (PEC) and a probabilistic species sensitivity distribution for the respective ENMs35. A procedure for calculating PECs of the investigated ENMs and assessing potential risks is shown in Fig. 1.

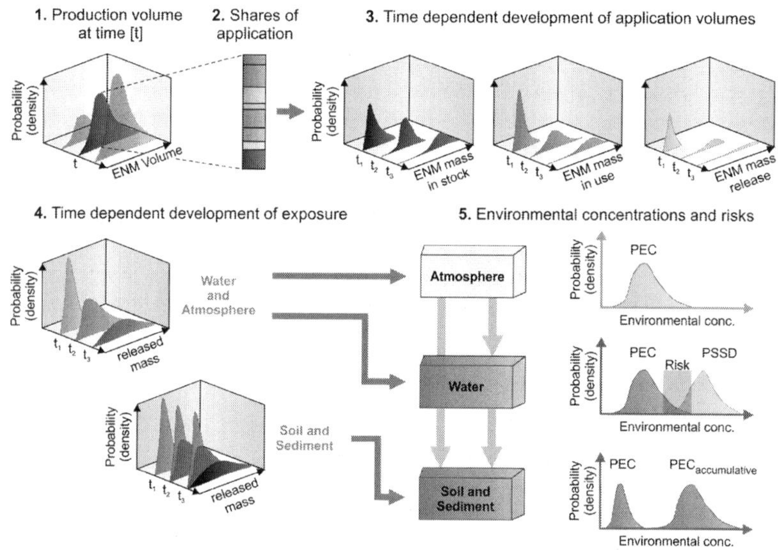

Figure 1: Data processing scheme for the calculation of a PEC and risk assessment. Each ENM was examined separately in the given order. After an initial estimation of the annual production volume of the respective ENM (1), according to the shares for different applications in processes and products, the mass fractions for these uses were calculated (2). Based on general trends for production volume, and taking into account factors for ENM release during its production, manufacture of ENM-containing products, use of ENMs and respective products, and end-of-life (EOL), the temporal distribution of ENM volumes for (i) stock, (ii) use, and (iii) release were determined stochastically (3). Information on release volumes enabled the computation of temporal probability distributions of ENM volumes in the atmosphere, water, and soil incl. sediment (4). To consider a large range of potential ENM concentrations for soil and sediment, we took into account two options for the persistence of ENM: on the one hand, an annual degradation of the released amount and, on the other hand, full ENM persistence and therefore an accumulation of annual releases that lead to considerably higher environmental concentrations of the respective ENM ($PEC_{accumulative}$). Besides the relevant volume within the environmental compartments, flux between compartments has to be considered for the calculation of PEC (gray arrows). Comparison of PEC with probabilistic species sensitivity distribution (PSSD) curves for the three investigated ENMs enabled a risk evaluation for waterbodies.

The ENM exposure modelling done so far is mostly grounded on mass flow and multimedia fate and transport analysis, as well as their combinations31.

Those models range from constant, annual release mass predictions on the continental or country level36 to dynamic, locally resolved computations with detailed ENM fate models integrating climatic and temporal variation31. We mainly focus on release and do not provide mechanistic fate modeling. ENM fate in natural and technical environments is covered as simplified mass transport reflecting erosion, sedimentation, dissolution, and other effects. However, this approach is far from proper mechanistic fate analysis, which is very challenging and is currently intensively researched and discussed37·38·39·40·41·42·43·44·45·46·47·48·49·50·51·52. Our modeling emerges from the initial stochastic probabilistic mass flow modeling36performed on several occasions for different ENMs and with mostly constant release, temporal, and geographic conditions53·54·55·56·57·58. Dynamic stochastic modeling was first developed for unilateral flows of ENM transport in incineration plants58. Similar attempts to analyze environmental release variation over time for ENM analysis of general environmental exposure have been done recently based on other models4·34·59·60. Several further ENM modeling studies tracing materials through economy into nature have been conducted in various other labs61·62·63·64·65. We go a step further in release estimations at the regional (country) level by fully dynamizing the ENM life cycle and end-of-life (EOL) input from ENM product use and disposal, as well as accounting for data that suggest ENM dispersion inside the target system. Nevertheless, our intercompartmental transport only reflects annual mass transfer rates at the country level; it does not mechanistically include aggregation, advection, aerosol attachment, suspension, resuspension, and other ENM fate processes, as has successfully been performed on local scales31. Three stochastic models cover (i) ENM release and subsequent transport between giving and receiving compartments, (ii) ENMs in economic circulation (delayed release), and (iii) ENMs already out there in nature and technical sinks over a large time period of up to one century (in extremes from 1950 to 2050). For the first time, we attempted to distinguish predicted concentrations derived from ENM use (USE) phase release, and potential after-use (EOL) release. Finally, we will scrutinize our predictive ENM modeling outputs by cautiously and self-critically discussing their validity and plausibility and comparing our predictions to results of first initial measurements.

RESULTS

ENM Production Volumes

Large variance in our survey and in the literature reflects extreme uncertainty regarding ENM production volumes in the nanomaterials community (Table 1). For SiO_2-ENM, estimates in the survey vary between a global production of more than 100,000 and roughly 3 million tons per year. Figures from literature begin far below 100,000 tons per year66 and range up to 1.5 million tons67.

Table 1: Estimates on global engineered nanomaterial production volumes from a survey and literature. For a detailed presentation of data from literature

Source	SiO_2-ENM in t/a	CeO_2-ENM in t/a	Ag-ENM in t/a
Company A	>1,000,000	10,000–100,000	1–10
Company B	ca. 3,000,000		
Company C	>100,000	1,000–10,000	10,000–100,000
Company D			1,000–10,000
Institute A	>100,000	1,000–10,000	100–1,000
Literature	55–1,500,000	5.5–10,000	5.5–550

Compared with numbers in the literature20·62·67 survey data for CeO_2-ENM reflect the upper limit of 10,000 tons per year, except for one of three estimates, which is higher by one order of magnitude.

For Ag-ENM, all data from the literature indicate a production volume below 1,000 tons per year2·20·21·56·66·67. Therefore, two of four estimates in our survey almost certainly represent extreme overestimates of the annual production and thus have been omitted for further modeling.

Estimates from our survey and data from literature were used to model a probability distribution of the global production volume. Regardless of their origin, all data from literature and our own survey where integrated with equivalent weighting. Before computing a distribution, data from literature were scaled for 2015 conditions using trends for the annual produced amount (see next section).

Trends for Annual Produced Amount of ENM

On reviewing our survey results and a recent project report for the European Commission5, we derived a median annual growth of 5% (with 50% uncertainty/variability on each side) for the time period after 2000, which we incorporated in our modeling approach. For individual engineered nanomaterial applications before 2000, a median annual growth of 1% was chosen based on application trends for Ag-ENM for medical purposes34 .

ENM Applications

Based on the evaluation of a combined investigation for applications of SiO_2-, CeO_2-, and Ag-ENMs, we identified product categories for each ENM.

Synthetic amorphous silica can be differentiated into pyrogenic or precipitated silica, silica gel, or sol; each has unique physical and chemical properties that may lead to different toxicological effects. Toxic effects and indications for inflammatory reactions are reported for pyrogenic (fumed) silica in cell culture assays68·69·70. For colloidal silica, only minimal overall toxicity was shown by Zhang, *et al*.68. Therefore, the different polymorphs of SiO_2-ENMs are treated separately in the present work. According to a study by the European Centre for Ecotoxicology and Toxicology of Chemicals (ECETOC) on SiO_2-ENMs9, pyrogenic silica is predominantly used in elastomers, polyester and epoxy resins, adhesives and sealants, and paints and coatings. Major applications of colloidal silica include the production of pulp and paper, refractories, and casting processes. Among numerous other applications, silica gel is primarily used as a food or tableting additive, as well as in paints, coatings, and plastics. The greatest proportion of precipitated silica is used for rubber products like tires and shoe soles9·57.

The investigation of product categories for CeO_2-ENMs revealed a number of different applications. Due to its oxygen storage capacity, CeO_2-ENMs are contained in the catalytic layer of ACCs and used as a fuel additive in diesel engines. Other applications include exterior coatings (e.g., UV-protective varnish for wood), integration in mixtures for CMP of glass and silicon wafers, and use in nickel–metal hydride (NiMH) batteries.

Although, for Ag-ENMs, absolute use volumes are low compared with SiO_2- and CeO_2-ENMs, Ag-ENMs are widely used. According to our investigation of its major product categories, it is used primarily in consumer electronics, in textiles, and for medical purposes.

ENM Releases

For each individual product category, a release dynamic was assumed using a transfer coefficient (TC) as a simplified representation of the release during (a) production, formulation, and manufacturing processes; (b) a product's use over its life span; and (c) end of life. TCs were mainly derived from data given in the literature. However, a complete list of the life cycle phases for a number of applications are not covered by existing publications on environmental release, such as EOL for use in SiO_2-ENM investment casting or CeO_2-ENM as a fuel additive. In such cases, TCs for EOL were obtained from literature focused on the respective processes or by primary research. There was not any information available regarding the use phase of some other applications. In these cases, assumptions were derived from environmental release categories (ERC) from the European Chemicals Agency (ECHA)71 or adopted from comparable releases of other materials. For example, TCs for CeO_2-ENMs used in chemical-mechanical-polishing (CMP) are based on ERC No. 12a, 'Processing of Articles at Industrial Sites with Low Release', and a report on recycling of rare earths72. Given the high number of different applications for the three ENMs investigated, a 'patchwork' approach in terms of TCs was inevitable due to the lack of available data. Corresponding uncertainties were taken into account by transferring TCs into probability distributions.

Overall, emissions in the first phase of the life cycle during production, formulation, and manufacturing processes are low and mainly associated with solid waste disposal, followed by emissions to waste water and a minor fraction to air. According to factors given in literature, release can be expected to not exceed 1% of the total ENM mass contained in most products, except for formulations of suspensions, where a release of 1–5% must be considered.

TCs for release during product life have been arranged separately for each product category. In accordance with the differentiation for individual types of synthetic amorphous silica, TCs of product applications have been treated separately. Relevant release sources for *pyrogenic silica* (due to share of product application and TC) are (a) non-silicones like pharmaceuticals, cosmetics, toner, batteries, or food with a major release to waste water and a release-enhancing product lifespan of only one year; and (b) silicones-elastomers and paints or coatings with a major release to soils. Although their release is distributed over a long product lifespan of approximately ten years, a high fraction of early release is assumed for paints and coatings.

For *colloidal silica*, excepting clarification processes of wine, beer, and fruit juice, which account for only about one-tenth of colloidal silica applications, we assume only very low release.

Important applications of *gel silica* are (a) pharmaceutical, food, and dentifrices; and (b) textiles with a high TC to waste water. However, for textiles, the release is stretched over an assumed medium product lifespan of five years. For absorbents and desiccants containing gel silica, a high release was assumed for all compartments, but they represent only a minor share with regard to product applications. Release-relevant applications of *precipitated silica* are (a) rubber products like tires and shoe soles, with a high TC to soils due to dust from streets; and (b) toothpaste and cosmetics with a major release to waste water.

Due to preliminary evidence for differential toxicological impacts of the different types of silica, it seems necessary to include, in a discussion on potential risks, an estimate of the contributions of silica types to overall release. Wastewater is an important compartment for release of silica during the use phase. According to its annual release masses into waste water, with 62% precipitated silica shows the highest mean annual contribution. In the outcome of our modeling approach, precipitated silica is followed by gel silica (22%), pyrogenic silica (12%) and colloidal silica (4%).

A relevant source of CeO_2-ENM emissions may result from (a) improper disposal of small NiMH batteries or (b) (in a worst-case scenario for fabrication conditions) from CMP during manufacturing of optical or electronical equipment. Although absolute numbers of TCs are lower, release from ACCs and diesel fuel additives represents a diffuse source whose potential for minimizing ENM emissions is smaller than for CMP or batteries, largely as a result of being contained within a diesel particulate filter or trap.

According to our investigations, most released nanosilver is derived from textiles during washing processes. Cleaning agents and cosmetics are also relevant sources of nanosilver, with a release path via waste water. Medical products, as the oldest application for nanosilver, contribute to release as well, with emphasis on emission to waste water.

As seen in Fig. 2, for CeO_2, release of ENMs to nature is dominated by release during the use cycle; however, the overall material transfer will be dominated by the flow into the technical compartments at the EOL phase. Assuming that we can reliably control most uncertainty and natural and economic variability in its main input parameters, the model (Fig. 2) effectively predicts the expected total mass transport and release spectrum. However, the prognostic precision disappears towards the middle of the century, where no clear tendency appears in the results. Based on the current data, the EOL release to nature is insignificant and smaller by a factor of 1,000 than the release from ENM product use. Table 2 confirms that, for all materials and time periods studied, release into nature is much lower than EOL mass transfers into technical sinks. Only silica reveals a significant discharge to nature, smaller only by a factor of approximately 2 than the mass ending up in technical sinks. This is explained by a large and diverse nano-silica application spectrum, with some applications leading to direct release into natural compartments during their use, like paints and coatings or rubber products with abrasive uses (e.g., tires or shoe soles). As seen in Fig. 3, the different nano-silica types vary considerably regarding their use-based release properties. Precipitated nano-silica contributes more than half of the waste water–based aquatic discharge. As we see in Fig. 4and Table 3, for the CeO_2 case, the different ENM applications considerably differ regarding their use life cycle and release to nature, as well as regarding their disposal and other EOL flows. From the CMP application, approximately 5% of mass may end up in natural environments, whereas, from ACCs, around 1% is released to nature. We learn that specific consideration of each application and its life cycle duration is necessary to significantly improve precision in this kind of release and subsequent exposure and risk assessment.

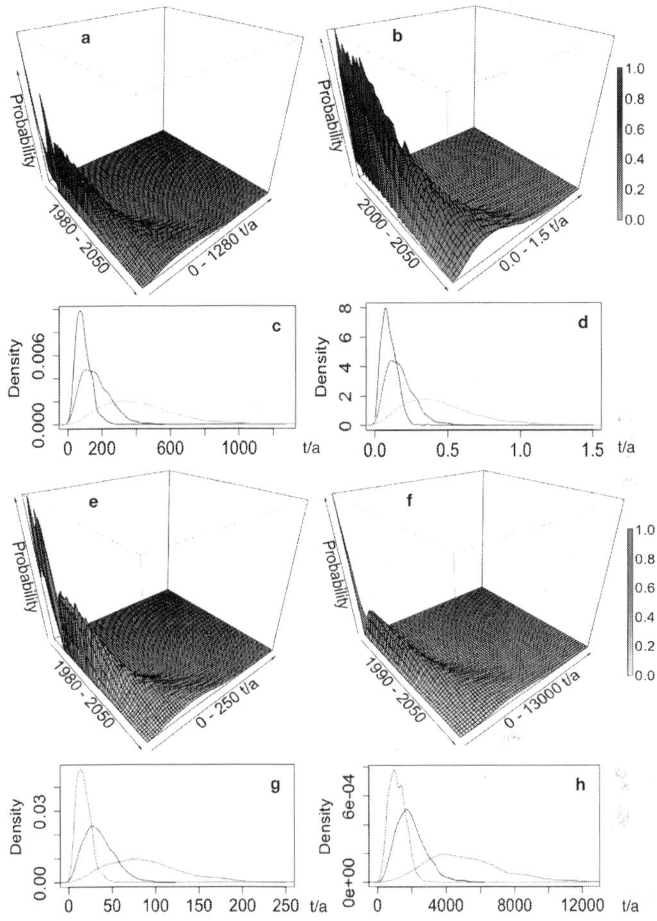

Figure 2: CeO_2-ENM released into natural (waters, soils, air) and technical compartments (landfills, recycling, waste incineration, and sewage treatment). The former release is shown in the Figures (a–d), the latter in the Figures (e–h). The left side (a,c,e,g) shows ENM released during product uses. The right side (b,d,f,h) represents release during ENM products' end-of-life phase. The corresponding relative density curves (c taken from a, d taken from b, gtaken from e, h taken from f) represent, from left to right, outputs for the years 2017, 2030, and 2050. Shown are the whole ranges (range with some probability) of the stochastic modeling. Scenarios that theoretically run all ENM applications simultaneously at the lowest, or, in another scenario, at the highest ENM production, use and environmental release are not considered here.

Table 2: Ag-, CeO_2-, and SiO_2-ENM release in t/a to nature (waters, soils, air) and the technosphere (landfills, recycling, waste incineration, and sewage treatment). Shown are rounded and simplified modal values (most frequent Monte Carlo outputs) for use and end of life (EOL) release. For CeO_2 **results, see also Fig. 2 and, for specific flow charts**

		2017		**2030**		**2050**	
		Use	**EOL**	**Use**	**EOL**	**Use**	**EOL**
Ag-ENM	*Nature*	<5	≪1	≈5	≪1	≈15	≪1
	Technosphere	<2	≈10	<5	≈15	<10	≈30
CeO_2-ENM	*Nature*	≈75	<1	≈ 100	<1	≈300	<1
	Technosphere	≈13	≈1000	≈30	≈1700	≈75	≈4000
SiO_2-ENM	*Nature*	≈17000	<5	≈30000	≈ 5	≈73000	≈15
	Technosphere	≈5250	≈36000	≈12000	≈55000	≈30000	≈145000

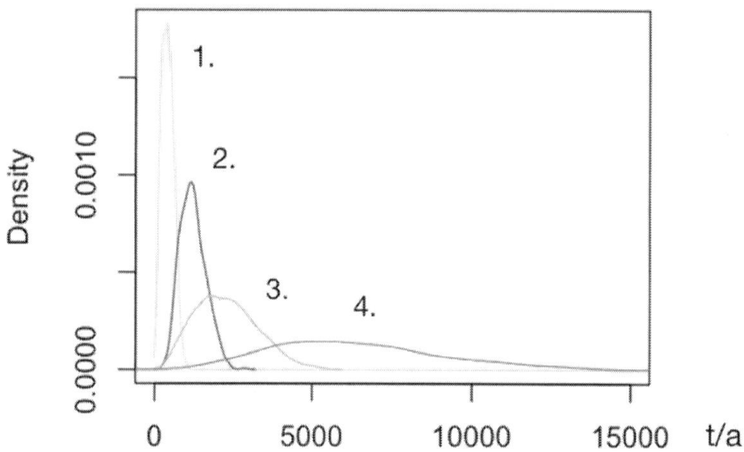

Figure 3: SiO_2-ENM contributions in t/a (*x* axis) to use release into wastewater in 2017 (most probable range). From left (mean contribution to total release): 1. colloidal silica (4%), 2. pyrogenic silica (12%), 3. gel silica (22%), and 4. precipitated silica (62%).

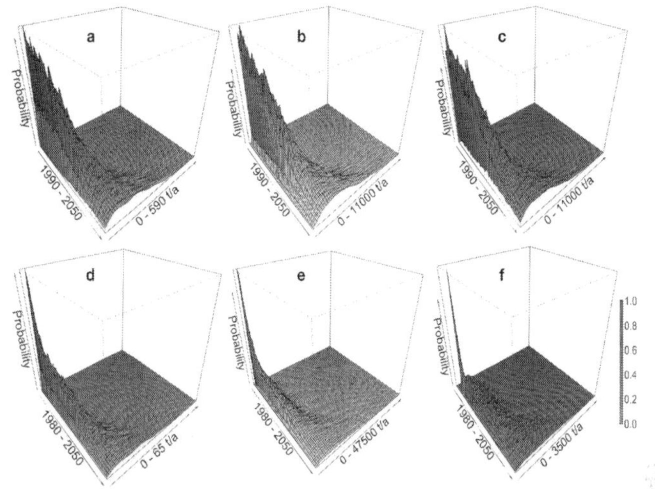

Figure 4: Predicted ENM mass (most probable range) between 1990 and 2050 (1980 and 2050) from two exemplary and relevant CeO_2-ENM applications. From left: ENM release mass to nature (a and d), ENM mass (b and e) in circulation (release delay during product use), and ENM mass (c and f) deposited or eliminated in technical treatment (landfilling, recycling, waste incineration, and sewage treatment). Above (a–c), chemical-mechanical polishing (CMP) application; below (d–f), use for ACCs (automotive catalytic converters).

Table 3: Predicted annual mass (most probable range) in tons (see corresponding Fig. 4) of CeO_2-engineered nanomaterial from use in single applications. CMP stands for chemical-mechanical polishing, ACC for automotive catalytic converters. In bold: mass in circulation (release delay during product use); italic: release mass to nature; and bold italic: ENM mass deposited or eliminated in technical treatment (landfilling, recycling, waste incineration, and sewage treatment)

Single ENM applications		2017			2030			2050		
		min	mode	max	min	mode	max	min	mode	max
In circulation	CMP	<3	mean* ≈600	≈3000	<5	≈1000	≈5000	<40	≈2000	≈11000
	ACC	<10	≈3000	≈10000	<20	mean ≈5000	≈20000	<200	mean ≈13000	50000
Release to nature	CMP	≈0	mean ≈30	≈120	<1	mean ≈60	≈250	<1	mean ≈150	≈600
	ACC	≈0	<2	≈10	≈0	≈2	≈20	<1	mean ≈15	≈70

Release to sinks	CMP	<5	mean ≈500	≈2500	<5	≈450	≈4000	≈5	mean ≈2850	≈11000
	ACC	<1	≈250	≈900	<1	mean ≈400	≈1500	<5	mean ≈1100	≈3500

ENM Concentrations

The model output reveals (as presented in Table 4, Figs 5 and 6) that it is impossible to establish any deterministic concentration prediction for ENMs. However, it is possible to show some probable tendencies and to explore the probable and possible range of concentrations. The full dynamic model output spectrum in one year and over the considered time periods (1950 to 2050) spans in extremis several orders of magnitude. The lowest aquatic ENM concentrations were predicted for silver and ranged (full spectrum) for fresh water in 2017 from a few pg/l to some ng/l. For 2030 and 2050, these predictions are, roughly speaking, about two- and sixfold higher, with most probable 2050 concentrations ranging from approximately 100 pg/l to 10 ng/l. The PECs for marine waters reveal, without exception, marginal values below pg/l levels and, in 2050, at the highest extremes, up to 1 pg/l. The fresh sediment concentrations range from a few pg/kg in 2017 (fast ENM degradation after one-year residence) to a few mg/kg in 2050 (no ENM degradation in nature). The marine sediment outputs are up to tenfold smaller and probably do not achieve mg/kg levels by 2050. In terrestrial contexts, the same time and scenario comparisons yield a range for agricultural soils between 30 pg/kg (min., 2017) up to 10 μg/kg soil (max., 2050). Similar outputs were seen for natural and urban soils. For sludge-treated soil areas, our predictions were about a factor of 30–40 higher than for agricultural soils. Equivalent time range predictions in the atmosphere (min. 2017 to max. 2050) span from low or below pg/m^3 to a few ng/m^3.

Table 4: PECs for 2017 for SiO_2–, CeO_2–, and Ag-ENMs. Shown are the most frequently modeled (mode) values; modes that only reflect concentrations from end of life (EOL) releases; ranges (range with some probability) of stochastic modeling that do not necessarily force all applications simultaneously running at the lowest, or, in another scenario, at the highest production; and use and environmental release levels. Such simultaneous extreme scenarios are provided in the absolute min-max borders on the left and right sides. The max border shows extreme high-end spectrum values, which are possible but not very likely, with the assumption that the entirety of ENM import and production (all applications) are simultaneously running at maximum levels with subsequent high ENM use and release. We have rounded the model output to two places after the decimal point. Nevertheless, the results included after the decimal point values should not suggest any accuracy of individual values taken from the analysis of probability distributions. This table rather shows the expected full spectrum with modeled trends

	Unit	**Min**	**Mode**	**(Mode EOL)**	**Range**	**Max**
SiO_2-ENM						
Sewage treatment effluent	**µg/L**	0.01	44.37	0.01	9.04–821.53	1,557.61
Surface water (fresh water)	**µg/L**	0.00	5.34	0.00	1.80–11.63	25.55
Sea water	ng/L	0.00	0.12	0.00	0.01–0.39	1.11
*Sediments (fresh water)**	mg/kg	0.00	32.12	0.00	0.15–131.13	302.82
*Sediments (fresh water)***	mg/kg	0.00	920.81	0.04	4.63–3,849.29	8,681.59
*Agricultural soils**	**µg/kg**	0.00	62.82	0.01	25.61–127.75	258.17
*Agricultural soils***	**µg/kg**	0.08	1,698.93	0.33	945.72–2,622.65	6,981.61
*Sludge treated soils**	**µg/kg**	0.11	3,085.15	0.42	1,125.36–6,508.83	11,603.86

*Sludge treated soils***	**µg/kg**	2.85	78,275.11	9.41	32,021.29–177,840.16	294,408.31
Air	ng/m³	0.00	13.92	0.10	1.89–51.36	79.43
CeO_2-ENM						
Sewage treatment effluent	ng/L	0.03	199.92	0.26	20.30–888.75	1,867.90
Surface water (fresh water)	ng/L	0.00	7.01	0.01	0.51–25.59	61.74
Sea water	ng/L	0.00	0.00	0.00	0.00–0.001	0.003
*Sediments (fresh water)**	**µg/kg**	0.00	49.03	0.02	0.34–253.49	611.95
*Sediments (fresh water)***	**µg/kg**	0.04	901.12	0.54	4.40–8,328.88	13,496.61
*Agricultural soils**	ng/kg	0.10	377.41	0.30	43.67–1,491.87	2,943.37
*Agricultural soils***	ng/kg	2.10	7,963.94	6.24	1,314.06–36,210.88	62,109.92
*Sludge treated soils**	ng/kg	0.97	4,450.17	6.65	528.57–19,011.99	35,069.36
*Sludge treated soils***	ng/kg	25.96	118,515.88	138.04	11,212.11–560,423.28	933,959.89
Air	ng/m³	0.00	1.23	0.00	0.12–5.53	12.10
Ag-ENM						
Sewage treatment effluent	ng/L	0.06	18.89	0.00	1.24–103.79	151.25
Surface water (fresh water)	ng/L	0.00	0.38	0.00	0.03–2.79	4.17
Sea water	ng/L	0.00	0.00	0.00	0.00–0.000	0.000
*Sediments (fresh water)**	**µg/kg**	0.00	1.85	0.00	0.02–33.67	47.02

*Sediments (fresh water)***	**µg/kg**	0.00	29.55	0.00	0.19–470.65	749.35
*Agricultural soils**	ng/kg	0.03	7.76	0.00	0.24–67.73	80.30
*Agricultural soils***	ng/kg	0.50	147.13	0.02	9.40–792.23	1,522.66
*Sludge treated soils**	ng/kg	1.13	257.18	0.01	20.08–1,661.16	2,294.05
*Sludge treated soils***	ng/kg	22.66	5,141.25	0.25	464.15–24,995.35	45,859.26
Air	ng/m^3	0.00	0.06	0.00	0.00–0.49	0.58

*100% degradation scenario after one year.

**100% persistent ENM (no-degradation scenario).

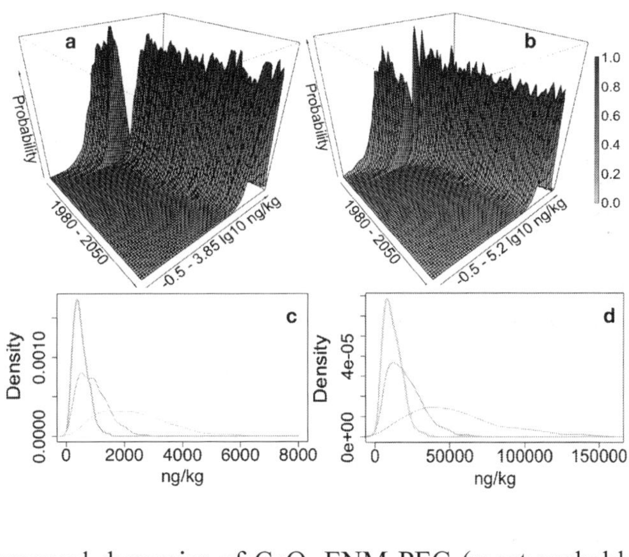

Figure 5: Temporal dynamics of CeO_2-ENM PEC (most probable range) for agricultural soils (left side, scenario of immediate ENM degradation after one year (a and c); right side (b and d), persistent scenario of no ENM degradation or elimination at all after being released into soils). The density curves refer in each scenario (c out of a, and d out of b) to the results for 2017, 2030, and 2050 (from left).

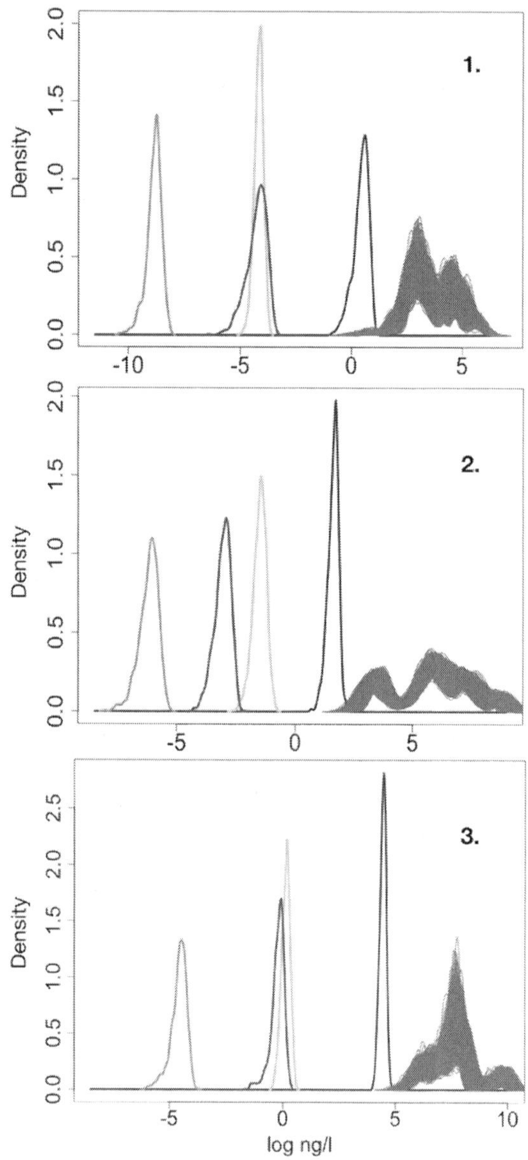

Figure 6: Water risk evaluations (1) for Ag-, (2) CeO_2-, and (3) SiO_2-ENMs by comparing PECs (most probable range) of 2050 and 1,000 PSSD curves (in red). The PEC curves (form left to right) represent marine water (green) loaded only from releases associated with end of life (EOL) of ENMs, in blue marine water loaded from ENM product uses, and EOL releases followed by the equivalent freshwater loadings from EOL (light blue) and total (use and EOL) release (dark blue).

The 2017 CeO_2-ENM full-spectrum concentrations predicted for fresh waters range from, minimally, 1 pg/l in 2017 to, maximally, a few hundred ng/l in 2050. The most probable current concentrations range approximately from 1 ng/l to 30 ng/l. Probable values for 2030 and 2050 are about 2–3 and 4–6 times higher, respectively. The maximal PECs for marine waters achieve, at the most, about 10 pg/l in 2050. The fresh sediment predictions span from fast ENM degradation values of a few ng/kg in 2017 to some mg/kg for the no-degradation scenario. The sea sediment equivalents are, on average and roughly, a factor 6–10 lower. The time and min-max scenario comparisons between 2017 and 2050 reveal, for agricultural soil PECs, a range from 100 pg/kg up to 400 μg/kg. Comparable outputs were again computed for natural and urban soils. The sludge-treated soil predictions were roughly a factor of 10–20 higher than those for agricultural soils. Our air PECs (min. 2017 to max. 2050) range approximately from a few pg/m^3 to about 50 ng/m^3.

The highest predictions concern the SiO_2-ENM concentrations. The fresh water extreme PECs range, in 2017, from pg/l levels to around 30 μg/l. The current most probable concentrations lie around 1–10 μg/l. In 2030 and 2050, these values are, again very roughly, 2 and 4–6 times higher, respectively. The marine water high-end extreme PECs are, in 2050, at the most around a few ng/l. Amorphous SiO_2-ENMs have a low water solubility9. It is therefore expected that the vast majority of released SiO_2-ENMs settle in soil and sediment. The ENM distributions obtained by our model for marine and fresh water represent the whole mass of the respective ENMs, including dissolved and particulate ENMs. The concentrations in fresh sediment span from about hundred ng/kg in 2017 (fast ENM degradation) to low g/kg levels in 2050 for the no-degradation scenario. The sediment results for marine contexts are, again, in most cases smaller by a factor of 6 or 7, probably achieving, at the most, a few g/kg in 2050. The time and scenario comparisons for agricultural soils show values lying between 3 ng/kg (min. 2017) up to 30 g/kg soil (max. 2050). As seen before, the natural and urban PECs were similar. The sludge-treated soil results were in most cases about 30–40 times higher than the agricultural equivalents. The full-spectrum time and min-max range predictions for atmospheric PECs (2017–2050) span from the before-computed low, below the pg/m^3 level, to some hundred ng/m^3.

ENM Risks

In Fig. 6 and Table 5, the risk probability plots for waters are shown. Those risks are predicted by contrasting our PECs for 2017, 2030, and 2050 to our probabilistic species sensitivity distribution (PSSD) modeling for the target materials. The risk computations involve concentrations in: (from right in Fig. 6) (i) freshwaters potentially contaminated by ENM product use release, combined with release from ENM products' EOL treatment; (ii) fresh water contaminated only by EOL-based release; (iii) marine water contaminated by use and EOL release; and iv) marine water contaminated by EOL release alone. An overlapping of those concentrations with the toxic limit curves reveals possible risks for ecosystems. The quantified risk percentages (Table 5) combine the probability of critical PECs, defined as values higher than the lowest PSSD, with the probability of the PSSD curve being below the maximal PEC. For the first half of this century, our results suggest, at most, low risks for natural waters loaded with the ENMs studied. EOL-based exposure and seawater concentrations seem, with some reservations as discussed at the end of this section and applying to all risk statements made here, so far harmless for all water organisms. For CeO_2-ENMs, risk is almost nonexistent. The Ag-ENM results for freshwater (use- and EOL-based ENM load) show risk values up to 1.4%, suggesting the possibility (at all time points studied) of exposure to a small fraction of water organisms. The same applies to SiO_2-ENMs, where, at least for 2050, significant exposure cannot be excluded. With the exception of Ag and SiO_2, the results show extremely low to no risk. However, a fly in the ointment is that, as seen in Fig. 6, by 2050 almost all Ag- and SiO_2-ENM PECs for fresh water are toxic for at least a small fraction of organisms. Another reason for caution is that our PEC simulations do not address short-term local extremes, which could lead to much higher loads. In extreme cases, Gottschalk, *et al.*54 found, for geographic and temporal variation influencing Swiss rivers, increase in ENM loads by factors up to the four- and five-digit range. Taking such extreme factors to hypothetically increase our freshwater total PECs would fully move the concentration curves of all target substances into the PSSD concentration levels (Fig. 6) resulting in all organisms being at risk.

Table 5: Risk results for Ag-, CeO_2-, and SiO_2-ENM in waters (see also Fig. 6). Risks are predicted by comparing PEC (most probable range) of 2017, 2030, and 2050 to PSSD outputs. The values represent fresh water loaded from ENM use and EOL releases, as well as fresh water loaded only from EOL release and marine water loaded from use and EOL, as well as only EOL, release. The percentages combine the probability of critical PECs (PEC values higher than the minimal PSSD) with the one of critical PSSDs (PSSDs smaller than the highest PEC)

Time	**2017**	**2030**	**2050**	**2017–2050**	**2017–2050**
Load	***Freshwaters with total ENM load***			**End of life (EOL) phase loaded Freshwater**	**Seawaters with total ENM load**
Ag-ENM	0.1%	0.7%	1.4%	0%	0%
CeO_2-ENM	0%	0%	≈0%	0%	0%
SiO_2-ENM	0%	0%	<0.1%	0%	0%

Model Results versus Measurements

The validation of modeled PECs is currently hampered by lack of reliable data on environmental concentration of the ENMs. Nevertheless, recently concluded monitoring projects[73,74] provided initial data on the occurrence of nanoscale CeO_2 and Ag particles in effluent waters of wastewater treatment plants and in surface waters (Table 6). In these studies, samples were collected from 25 measurement points within the Bavarian survey network of watercourses. Amongst others, the authors of these reports used asymmetric flow field fractionation (AF4) in combination with (single particle) inductively coupled plasma mass spectrometry (ICP-MS)[73] to detect and quantify nanoscale Ce- and Ag-particles. The target particles were detected in both, industrial waste waters and river waters. The reported concentrations have to be interpreted as maximum values as they refer to the total nanoscale particles detected in the respective samples and include natural, incidental and engineered NP. Nevertheless, reported concentrations of selected nanoscale particles are in good agreement with the respective modeled concentrations. The maximum value of the measured concentrations for nanoscale Ce particles is slightly higher than the respective modeled concentration of CeO_2 nanoparticles in wastewater treatment plant effluents. For the Ag-NP, maximum measured concentrations are higher by about one order of magnitude compared to the modeled concentrations. Considering that very high effluent concentrations are probably caused by industrial wastewater treatment plants (and such spatial heterogeneities are not covered by the model) and that the model only includes engineered NP, the agreement between the modeled and the measured concentrations is surprisingly good. In the case of non-diluted waters, concentrations of silver – NP in our generic wastewater treatment plant effluents are in the same order of magnitude (ng/L) as the nanoscale silver reported from industrial wastewaters.

Table 6: Comparison for an early validity discussion of our PECs and nanoparticle concentration measurements, reported elsewhere73. We show full ranges with mode or mean values

Compartment Unit		*Modelled ENM*			*Measured nanoparticles*	
		Mode	**Range**	**Mean**	**Range**	
CeO_2-ENM	*Sewage treatment effluent**	ng/L	199.90	0.03–1868.00	na	0.00–3900**
	Surface water (fresh water)	ng/L	7.00	0.00–61.70	5.10	0.00–13.80
Ag-ENM	*Sewage treatment effluent**	ng/L	18.90	0.10–151.20	na	0.00–1300.00
	Surface water (fresh water)	ng/L	0.40	0.00–4.20	0.74	0.00–6.20

For CeO_2, Park, *et al.*17 measured up to 0.5 ng/m^3 near a busy road often frequented by Envirox-equipped busses. The authors also predicted, by assuming no sedimentation, 20–80 ng/m^3 for a busy street canyon and 0.3–1 µg/g of soil 96 m from a road. These soil PECs are in the same range and about a factor of 10 higher compared with our lower values for presumably less exposed agricultural soils: 0.06 µg/g of soil (persistent scenario). Our most probable air PEC range of 0.1–5.5 ng/m^{3}is in the same order of magnitude as the concentrations reported by the Park *et al.*17 and correlate fairly well with their predictions for a busy street canyon. Furthermore, Erdakos, *et al.*75 confirmed such concentrations, predicting up to 20 ng/m^3 for the United States.

DISCUSSION

The release and persistence of ENM into the environment was first detected a decade ago76. The investigation of production volumes of ENMs is of particular importance for the outcome of subsequent steps in exposure modeling. Our survey yielded numbers in the upper range of recent estimates from literature. However, there are indeed some indications for an increased production of ENMs. In the case of silica, leading manufacturers announced an increased productive capacity in recent press releases10·11. For CeO_2-ENMs, we calculated the annual amount of ENMs needed for products and processes independent of estimates on production volumes: with a sum of approximately 22,000 t/a, the result of our demand-based calculation confirms the outcome of our survey on annual production volume. In the case of nanosilver, the large variance of volumes in our survey reveals a certain amount of ignorance regarding production volumes among the experts contacted in our survey. We expect an annual production volume below 1,000 t for nanosilver.

As important diffuse sources for the release of ENMs with no possibility for removal or separation of nanomaterials (e.g., in sewage treatment plants), we identified CeO_2 in automotive applications and SiO_2 in tires and shoe soles.

Besides CeO_2-ENMs emitted from automotive exhaust systems, the majority of CeO_2-ENMs resides within diesel particulate filters or the catalytic converters. In Germany, both elements of the exhaust system are shredded. A minor amount of the scrap is melted in a plasma furnace, where Ce remains bound in the slag because of its high oxygen affinity. A share of German scrap from automotive exhaust systems is exported and sold to international smelters (Source: own research and personal communication with scrap processing companies). Owing to the fact that the fraction of scrap from exhaust systems treated within Germany is unknown, as a simplifying assumption, the entire mass of residual CeO_2-ENMs in exhaust systems is considered as a deposit to landfills.

According to Wang, *et al.*57, a large proportion of used tires is treated in waste incineration plants, including thermal utilization in cement production. Nevertheless, during use, one-tenth is released to soils. Whether SiO_2-ENMs are released from tires and shoe soles mainly as single particles, agglomerates, or as a composite of rubber and ENM including other components like carbon black particles, is not known. Unless detailed information on the condition of emitted silica particles from rubber products becomes available, they have to be considered as single nanoparticles.

As mentioned above, owing to their varying toxicological effects, a differentiation between the types of silica is advisable. For use-based release into waste water, our modeling approach assigns the major contribution to precipitated silica. This dominant role of precipitated silica is attributable to the fact that it accounts for nearly three-quarters of the overall amount of produced silica (see Table S3) and has a number of applications with relevant releases into waste water (see Table S7). Pyrogenic silica, a form of silica with reported toxicological relevance, accounts only for a minor part of silica release into wastewater. Therefore, in an evaluation of associated risks (see below), we have to be aware that potential toxicologically relevant forms of silica may represent only a part of the total amount of engineered silica in particular compartments, at least for waterbodies and sediments. Moreover, it becomes clear that further work on exposure to ENMs has to consider the different applications of ENMs in more detail.

The broad range of our stochastic model outputs (Fig. 2) reflects uncertainty propagation due to mixing high uncertainty and variability of economic (ENM use) and scientific (release) data. Years of research to improve ENM use-release kinetics will be necessary to exceed the threshold of current preliminary and assumptive release data and subsequent risk modeling that only reflects homogeneous ENM dispersal in the environment at regional to continental levels. Normalizing mass flows to concentrations using total countrywide volumes has led to the impression that significant dilution occurs everywhere and without exception. This assumption may be applied in making preliminary and general predictions. However, it has significant limitations for understanding local, short-term, high-extreme exposure scenarios31·54. For example, considering CeO_2 used for CMP implicates chemical and microelectronics industries near Dresden, which are on the Elbe River. Presumably, exposure in that region would be plausible in air, water, and agriculture. However, the residents of Hesse living along the Main might only experience air exposure caused by emissions from cars. In the United States, there is a mixing zone at the outfalls of all production plants. The U.S. Environmental Protection Agency (EPA) assumes a tenfold dilution and requests ecotoxicity data that are calculated at that concentration. These data must be available in order for the plants to receive effluent permits. We would have to quote that as the maximum concentration and then calculate the decrease in concentration caused by hetero-aggregation. This calculation would be similar to TiO_2 on the Rhine37 or to Ag and ZnO nanoparticles modeling on the James River40.

The initial model validation, or, more realistically, the model-to-measurement comparisons shown above, is challenging. The possibility of statistical Type-III error cannot be fully excluded, owing to possibly having stochastically combined mostly non-validated input data to generate acceptable model output results. Furthermore, the available measurements do not distinguish between natural and engineered nanoparticles. Thus, the correlation between and the plausibility of the measurements and model results can be found by comparing the maximum upper limits of the modeling values, which should not exceed the measured equivalents. However, this correlation may also be found since wastewater should be dominated by engineered (non-natural) nanomaterial.

Comparing various modeling results has, in contrast to the comparison of measurements with modeling outputs, very low or no mutual validation power. Nevertheless, some limitations and plausibility may be discussed by opposing a non-comprehensive selection of other model results in Fig. 7 to our predictions for current concentrations in fresh water and soils. It is striking that we present the widest range of results, which speaks for our rigorous uncertainty treatment and integration of several years of dynamic release. As further evidence of this, previous results with similar model output ranges34, even if not showing the full range of Monte Carlo simulations, cover (albeit only in a rudimentary way) the full uncertainty found in the raw data currently available for time-dynamic uncertainty propagation. Our silver predictions fully agree with existing observations for waters and soils. Although we completely overhaul former modelling approaches that ignored time-dependent release variation and uncertainty, and although we use completely different data for different regions, our output ranges agree with values already published in the last decade36. This also reflects the short ENM residence time in water (lower than one year), which makes a dynamization for such receiving compartments almost superfluous. High agreement is also seen for CeO_2 and SiO_2 results, despite their being based on or reflective of various data, models, and investigated regions. A relatively large water volume and a lower ENM use leads to lower Danish CeO_2-ENM water PECs77 that nevertheless fully lie in our range. Others39 have shown, for Europe, water PECs within our German range. However, the higher maximal values are not discussed in detail here because they partially exceed the relatively high concentrations measured in industrial wastewaters. Agreement is shown for soils, with the exception of high values reported in Meesters, *et al.*39. The silica-specific comparison with Wang, *et al.*57 shows good correlation; however, the reported range is narrower than our predicted range, owing to treating uncertainty in decades-spanning data less comprehensively as we do in our dynamic release modeling.

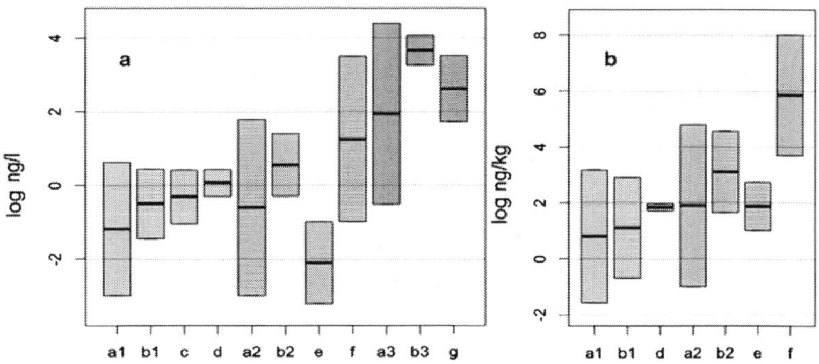

Figure 7: Comparisons of modeled PEC ranges (and means) for current values, logarithmically shown on the *y* axis for surface waters (Figure a) and soils (Figure b) (blue: Ag-ENMs; green: CeO_2-ENMs; orange: SiO_2-ENMs). Shown for all three materials are (a1–a3) the full range from our work, (b1–b3) our most probable range, (c) the full range of all PECs from the first probabilistic models36, (d) recent results34, (e) Denmark-specific PECs77, (f) results of a recent multimedia fate model39, and (g) silica-specific results57. Background colors are randomly chosen for clarity.

Modeling remains indispensable as long as analytical methods cannot distinguish background and engineered material. Preliminary environmental release, exposure, and risk assessment still leave room for at least partially misleading or erroneous findings. Thus, future preliminary risk estimation will be focusing on filling significant data gaps on ENM use and release amounts and overcoming significant modeling simplifications of non-mechanistic environmental fate analysis (dissolution, aggregation, sedimentation and other).

Closing the gap is a challenge. Industry has a reluctance to provide information for both competitive and anti-trust reasons, but also wishes to have ecotoxicity testing done at relevant exposure levels. The absence of this information has led some countries in Europe to institute chemical registries, which may broadcast production volumes of suppliers in those countries to their competitors. Some method of aggregating product statistics, perhaps monitored by trade associations and government agencies is yet to be found. Industrial data are needed on precise, current ENM mass production and use. Such data can be fed into more sophisticated economic nanoproduct consumption models. Even if such data become available, however, future ENM market dynamics will be difficult to predict, owing to their dependence on yet-unknown economic trends and regulative interventions4. Refined

ENM use models should be linked to robust nanoproducts experimental evidence and should be based on the most relevant available environmental release paths, kinetics, and material forms. Release from nanoproducts will have to account for a wide material spectrum ranging from pristine particles to fragmented products6.

Finally, uniformly distributed mass flows in target systems on the country level only rudimentarily represent nanomaterial fate in the environment. In particular, spikes in daily and other short-term and locally specific exposure and associated potential risks are not covered by simplified, annually resolved computations with ENM distributions homogenized to the country level. Garner, *et al.*31 refined multi-media fate modeling by accounting for locally and time-resolved intercompartmental ENM transfers and transformations distinguishing free, homo- and heteroaggregated, as well as dissolved, nanomaterial. The authors point the way to holistically and mechanistically considering the most significant processes. Those include, among others, ENM attachment on aerosols, advection and sedimentation in air, erosion and runoff during rain events, dissolution and heteroaggregation, suspensions, resuspension and burial in waters and sediments, and water and sediment flow into lakes and sea water.

A remaining question is to what extent ENM ecotox and subsequent environmental risk modeling is impaired by the use of inconsistent toxicity data. The challenge regarding highly uncertain or contradictory data for nanotoxic endpoints has recently been reviewed32. The most critical issues were shown to be the absence of nano-specific effects that may either not have existed or that may have been masked by practical challenges to test performance on the nanoscale, such as "high-doses" material (nanomaterial) that was not fully characterized. Last but not least, an extreme variation of test methods was found, combined with unknown or unreported nanomaterial properties (particle, size distribution, aggregation dynamics, etc.) and missing reference or control samples for the performed tests. While such critical review made many valid points, it did not address the uncertainty when 'official testing' requirements, those protocols established by the OECD or national bodies, actually require modification and lead to situations where nominally thorough testing can still be incomplete.

METHODS

Materials and Data

Based on a standardized questionnaire, a survey was conducted to assess the annually produced volume of SiO_2-, CeO_2-, and Ag-ENMs and their annual trends. 72 producers, dealers, and research institutions in the field of nanomaterials were contacted in the course of the survey. Despite repeated inquiry, useful information was hard to obtain. Of the 28 replying addressees, 20 refused to respond, either because of ignorance with respect to production volumes and trends or because of a restrictive company policy that prohibits giving any information or ratings in surveys. Eight questionnaires were completed. Six of these eight estimates turned out to be useful for further evaluation. One of the six remaining addressees is a research institution; the other respondents are trading companies or producers of ENMs.

The share of the German annual market volume of the investigated ENMs was calculated by multiplying the global annual production of the respective ENMs by the ratio of German annual domestic demand and the worldwide gross domestic product (GDP) (in market prices). Using the German domestic demand instead of the German GDP corrects influences of annual export and import volumes.

For the derivation of mass fractions according to product categories of the examined ENMs. Estimations for the start of use of ENMs are based on the results of a literature search.

Assumptions are based on results of our own investigations, data given in recent publications, databases of the German Department of Statistics, and ECHA's (European Chemicals Agency) environmental release categories78.

Data on German geographic and aquatic conditions, as well as statistical information on the German waste management system.

Release Modeling for use and End-of-life Material Lifecycle Phases

In order to provide comprehensive insight into the possible environmental exposure to the target ENMs, a full and time-dynamic lifecycle release spectrum from all currently known significant ENM use categories was modeled. As shown in Fig. 1, such modeling includes annual ENM mass input into the target system from domestic production or import of pure ENMs or products containing ENMs. This input is followed by mass flows into lifecycle phases of ENM product use, with or without ENM release to natural or technical environments. From such use phases, lifecycle time dependent ENM release to waters, wastewaters (and sewage treatment), soils, and air is modeled. Non-released ENMs at the end of such product use phases become EOL material treated in waste incineration, landfill, or recycling processes. Thus, the target system is never in equilibrium but constantly changes. A comprehensive inventory of the used ENM mass input and release coefficient parameters. Such dynamic ENM release was modeled for the time period 1950 to 2050. Still, most ENM use started at the end of the last or in this century. For all use categories, of the receiving compartments to which release of ENMs takes place. Such release is shown by equation 1for a first scenario with short residence time of the emitted ENM.

$$r_1(t) = \sum_i \sum_j I_i(t_j) \tag{1}$$

The release r_1 reflects the modeled release at time t as a material input (I) function of all ENM-releasing engineered nanomaterial applications i and their life cycle years j contributing to such release. The life cycle length m (equations 3–6) reflects the whole release period. Such release leads to non-cumulative release and exposure of ENM in waters, air, sediments, and soils characterized by a short residence time (up to one year) before being degraded, eliminated, or transformed into other compounds. In contrast to this first scenario of short residence times of ENMs in nature, we also modeled (equation 2) an opposite scenario on the other side of the fate spectrum, with ENM persistently remaining in soils and sediments with no degradation.

$$r_2(t) = \sum_s r_1(t_s) \tag{2}$$

This contrasting scenario reflects the same model input as given in the first scenario, but this time all computed material release (inputs [I]) from previous years are considered as well. That is of importance when the already released potential contaminant persistently remains in a particular natural or technical environment.

The *R*-based79 Monte Carlo implementation of such time-dynamic mass inputs and stochastic release from various applications into different ENM-receiving compartments is given in Fig. S1 and shown in detail elsewhere58.

It allows stochastic exploration for all parameters of a wide range of possible values throughout the whole target ENM transfer system in time and space dimensions. Such transfer reflects unidirectional ENM transport, as done on other occasions, based on probabilistic material flow analysis designed for complex multidirectional transfer systems36·80.

Use Release

The use-release model covers release during the whole ENM lifecycle before ending up in EOL treatment such as waste disposal, recycling, or landfilling. Hence, such use-phase release occurs during the nanomaterial engineering and manufacturing of products (including import of material and products) and the consumption phase of those products. This model considers, for each application (product) as shown in equation 3, the annual release at a particular time (particular year during the whole lifecycle of a target ENM application) by considering all release contributions coming from all relevant lifecycle years up to that time point.

$$r_{use}(t) = \sum_{j} \frac{I(t_j)}{m} \tag{3}$$

Equation 3 is derived from equation 1 (equivalent derivations for equation 2 are not shown) and didactically conceptualizes a regular (annual) time-dynamic release input for one single product use at time t for all its release-contributing product lifecycle years j up to time point t with m as lifecycle length of the corresponding product.

An additional use model reflecting the time-dynamic input into ENMs in circulation during use phases without any release (delayed release) has been modeled as well (equation 4). Such material in circulation mass covers the not-yet-released material at a certain time. This model considers all lifecycle years j not yet contributing to ENM release.

$$r_{delayed(use)}(t) = \sum_{j} I(t_j) \cdot \frac{(m + j - t - 1)}{m} \tag{4}$$

In addition to those use-release and use-in-circulation computations, we also modeled material no longer in consumption and release processes but already in nature (or in the technical sinks after EOL treatment (disposal, recycling, landfilling)). Such modeling may be derived from the release models by considering the past release periods or computed with their own input functions.

End of Life Release

EOL release covers, for each particular application, the ENMs not emitted during its whole product use life. That kind of release occurs, as shown in equation 5, at the end of the product life time *m*.

$$r_{eol}(t) = I(t_{t-m}) \quad (\mathrm{t} > \mathrm{m}) \tag{5}$$

For the EOL release modeling, the material in circulation (equation 6) refers to the ENMs in products not ending up in use-based release and waiting for EOL treatment via incineration, landfill, or recycling.

$$r_{delayed(eol)}(t) = \sum_{j} I(t_j) \; (\mathrm{t} \leq \mathrm{m}); \tag{6}$$

$$r_{delayed(eol)}(t) = \sum_{j=t-m+1} I(t_j) \; (\mathrm{m} < \mathrm{t} \leq 2\,\mathrm{m}) \tag{7}$$

Additionally, in the case of EOL modeling, ENMs already released (into nature or technical sinks) can be derived from the release computations by considering past release periods or from their own input functions.

Ecotoxicological and Risk Estimations

Risk has been estimated by comparing our environmental predictions to ecotoxicologically significant concentrations that can be derived from species sensitivity distribution (SSD) modeling. For water environments, some data were available for conducting PSSD analysis35, an approach allowing integration of highly controversial ecotoxicologial data for ENMs27·57·81. For the terrestrial compartments, PSSD could not be performed, because the data are too scarce and are even more uncertain than those for aquatic compartments.

To cope with the challenges of deriving and handling highly uncertain and controversial toxic endpoints, the Monte Carlo–based PSSD approach was used35. Such modeling does not eliminate uncertainties, but it stochastically explores a large spectrum of possible target toxic endpoints. In contrast to SSD methods that use a mean endpoint value for each single species, our PSSD develops a unique, single-species sensitivity distribution for each tested organism. If taken together, from all single-species distributions of all other tested organisms, the generic PSSD for a target ecosystem emerges.

For the Ag- and CeO_2-ENMs, we based our toxicological modeling on the endpoints taken from Holten Luetzhøft, *et al.*82, in the "Values" column of their Tables 17and 21. For the SiO_2-ENMs, the no-effect concentration endpoints are derived from Wang, *et al.*57, considering their "Concentration" column of the Table S7.

We applied a 50% uncertainty/variability interval around all endpoints fed into the ecotoxicology model. The same uncertainty/variability interval was applied around the factors (uncertainty factors 10) used to transform acute into chronic effects and the observed effects into no-effect predictions.

ACKNOWLEDGEMENTS

We are grateful to Martin Missong for support in calculating the German market volume. We thank Bernd Nowack and Andreas Gondikas for their helpful review and suggestions on our initial raw data evaluation. This work has been funded by the German Ministry for Science and Research (BMBF) within the study "Design criteria for sustainable nanomaterials (DENANA)" under code 03X0152A.

REFERENCES

1. The Royal Society & The Royal Academy of Engineering. *Nanoscience and nanotechnologies: opportunities and uncertainties*. (Royal Society, 2004).
2. Pulit-Prociak, J. & Banach, M. Silver nanoparticles – a material of the future…? *Open Chemistry* 14, 76–91, https://doi.org/10.1515/chem-2016-0005 (2016).
3. D'Mello, S. R. *et al.* The evolving landscape of drug products containing nanomaterials in the United States. *Nat Nano* 12, 523–529, https://doi.org/10.1038/nnano.2017.67 (2017).
4. Sun, T. Y. *et al.* Envisioning Nano Release Dynamics in a Changing World: Using Dynamic Probabilistic Modeling to Assess Future Environmental Emissions of Engineered Nanomaterials. *Environ. Sci. Technol.* 51, 2854–2863, https://doi.org/10.1021/acs.est.6b05702 (2017).
5. Ricardo Energy & Environment. Support for 3rd regulatory review on nanomaterials: Interim/Background Report. (2016).
6. Nowack, B. *et al.* Meeting the Needs for Released Nanomaterials Required for Further Testing—The SUN Approach. *Environ. Sci. Technol.* 50, 2747–2753, https://doi.org/10.1021/acs.est.5b04472(2016).
7. Bettini, S. *et al.* Food-grade TiO2 impairs intestinal and systemic immune homeostasis, initiates preneoplastic lesions and promotes aberrant crypt development in the rat colon. *Sci Rep* 7, 40373, https://doi.org/10.1038/srep40373 (2017).
8. Kahru, A. & Ivask, A. Mapping the dawn of nanoecotoxicological research. *Acc Chem Res* 46, 823–833, https://doi.org/10.1021/ar3000212 (2013).
9. JACC. Synthetic Amorphous Silica. (Brussels, Belgium, 2006).
10. Evonik. Evonik plans to build a new production plant for precipitated silica in the United States. (2015).
11. Solvay. Solvay to build a new silica plant in Poland (2013).
12. Fung, M. & Bowen, D. Silver products for medical indications: risk-benefit assessment. *J Toxicol Clin Toxicol.* 34, 119–126 (1996).
13. Nowack, B. Nanosilver Revisited Downstream. *Science* 330, 1054–1055, https://doi.org/10.1126/science.1198074 (2010).
14. Fabrega, J., Luoma, S. N., Tyler, C. R., Galloway, T. S. & Lead,

J. R. Silver nanoparticles: behaviour and effects in the aquatic environment. *Environ Int* 37, 517–531, https://doi.org/10.1016/j.envint.2010.10.012 (2011).

15. Hansen, S. F. & Baun, A. When enough is enough. *Nature Nanotechnology* 7, 409–411, https://doi.org/10.1038/nnano.2012.115 (2012).
16. Dahle, J. T. & Arai, Y. Environmental Geochemistry of Cerium: Applications and Toxicology of Cerium Oxide Nanoparticles. *Int. J. Environ. Res. Public Health* 12, 1253–1278, https://doi.org/10.3390/ijerph120201253 (2015).
17. Park, B. *et al.* Hazard and risk assessment of a nanoparticulate cerium oxide-based diesel fuel additive - A case study. *Inhalation Toxicology* 20, 547–566, https://doi.org/10.1080/08958370801915309 (2008).
18. Gottschalk, F., Sun, T. Y. & Nowack, B. Environmental concentrations of engineered nanomaterials: Review of modeling and analytical studies. *Environ. Pollut.* 181, 287–300 (2013).
19. Hunt, G. *et al.* Towards a Consensus View on Understanding Nanomaterials Hazards and Managing Exposure: Knowledge Gaps and Recommendations. *Materials* 6, 1090 (2013).
20. Lazareva, A. & Keller, A. A. Estimating Potential Life Cycle Releases of Engineered Nanomaterials from Wastewater Treatment Plants. *Acs Sustainable Chemistry & Engineering* 2, 1656–1665, https://doi.org/10.1021/sc500121w (2014).
21. Burkhardt, M., Englert, A., Iten, R. & Schärer, S. Entsorgung nanosilberhaltiger Abfälle in der Textilindustrie - Massenflüsse und Behandlungsverfahren. (HSR Hochschule für Technik, Rapperswil, Schweiz, 2011).
22. Nowack, B. *et al.* Potential scenarios for nanomaterial release and subsequent alteration in the environment. *Environ. Toxicol. Chem.* 31, 50–59, https://doi.org/10.1002/etc.726 (2012).
23. von der Kammer, F. *et al.* Analysis of engineered nanomaterials in complex matrices (environment and biota): General considerations and conceptual case studies. *Environ. Toxicol. Chem.* 31, 32–49, https://doi.org/10.1002/etc.723 (2012).
24. Walser, T. *et al.* Persistence of engineered nanoparticles in a municipal solid-waste incineration plant. *Nature Nanotechnology* 7, 520–524, https://doi.org/10.1038/nnano.2012.64 (2012).

25. Wigger, H. Environmental Release of and Exposure to Iron Oxide and Silver Nanoparticles - Prospective Estimations Based on Product Application Scenarios. *Wiesbaden, Germany, Springer Vieweg* (2017).
26. von Gleich, A., Steinfeldt, M. & Petschow, U. A suggested three-tiered approach to assessing the implications of nanotechnology and influencing its development. *J. Cleaner Produc.* 16, 899–909, https://doi.org/10.1016/j.jclepro.2007.04.017 (2008).
27. Gottschalk, F., Kost, E. & Nowack, B. Engineered nanomaterials in waters and soils: A risk quantification based on probabilistic exposure and effect modeling. *Environmental Toxicology & Chemistry* 32, 1278–1287 (2013).
28. Holden, P. A. *et al.* Evaluation of Exposure Concentrations Used in Assessing Manufactured Nanomaterial Environmental Hazards: Are They Relevant? *Environmental Science & Technology* 48, 10541–10551, https://doi.org/10.1021/es502440s (2014).
29. Uskali, M. Modelling Failure. In *Logic, Methodology and Philosophy of Science - Proceedings of the 15th International Congress (Helsinki). Ilkka Niiniluoto, Päivi Seppälä, Elliott Sober (eds)*. College Publications, UK. (2017).
30. Mäki, U. Models and the locus of their truth. *Synthese* 180, 47–63 (2011).
31. Garner, K. L., Suh, S. & Keller, A. A. Assessing the Risk of Engineered Nanomaterials in the Environment: Development and Application of the nanoFate Model. *Environ. Sci. Technol.* 51, 5541–5551 (2017).
32. Krug, H. Nanosafety research - are we on the right track? Some thoughts based on a comprehensive literature review. *Angew. Chem. Int. Ed.* (2014).
33. Nanowerk. Nanowerk-Database, http://www.nanowerk.com/(2017).
34. Sun, T. Y., Bornhöft, N. A., Hungerbühler, K. & Nowack, B. Dynamic Probabilistic Modeling of Environmental Emissions of Engineered Nanomaterials. *Environ. Sci. Technol.* 50, 4701–4711 (2016).
35. Gottschalk, F. & Nowack, B. A probabilistic method for species sensitivity distributions taking into account the inherent uncertainty and variability of effects to estimate environmental risk. *Integrated Environmental Assessment and Management* 9, 79–86 (2013).
36. Gottschalk, F., Sonderer, T., Scholz, R. W. & Nowack, B. Modeled environmental concentrations of engineered nanomaterials (TiO2,

ZnO, Ag, CNT, fullerenes) for different regions. *Environmental Science and Technololgy* 43, 9216–9222 (2009).

37. Praetorius, A., Scheringer, M. & Hungerbuehler, K. Development of environmental fate models for engineered nanoparticles - a case study of TiO2 nanoparticles in the Rhine River. *Environ Sci Technol.* 46, 6705–6713 (2012).
38. Liu, H. H. & Cohen, Y. Multimedia Environmental Distribution of Engineered Nanomaterials. *Environ. Sci. Technol.* 48, 3281–3292, https://doi.org/10.1021/es405132z (2014).
39. Meesters, J. A. J., Quik, J. T. K., Koelmans, A. A., Hendriks, A. J. & van de Meent, D. Multimedia environmental fate and speciation of engineered nanoparticles: a probabilistic modeling approach. *Environ. Sci.: Nano* 3, 715–727 (2016).
40. Dale, A. L., Lowry, G. V. & Casman, E. C. Stream Dynamics and Chemical Transformations Control the Environmental Fate of Silver and Zinc Oxide Nanoparticles in a Watershed-Scale Model. *Environ. Sci. Technol.* 49, 7285–7293 (2015).
41. Dale, A. L. *et al.* Modeling Nanomaterial Environmental Fate in Aquatic Systems. *Environ. Sci. Technol.* 49, 2587–2593, https://doi.org/10.1021/es505076w (2015).
42. Sani-Kast, N. *et al.* Addressing the complexity of water chemistry in environmental fate modeling for engineered nanoparticles. *Sci. Total Environ.* 535, 150–159 (2015).
43. Markus, A. A., Parsons, J. R., Roex, E. W. M., de Voogt, P. & Laane, R. W. P. M. Modeling aggregation and sedimentation of nanoparticles in the aquatic environment. *Sci. Total Environ.* 506–507, 323–329 (2015).
44. Quik, J. T. K., de Klein, J. J. M. & Koelmans, A. A. Spatially explicit fate modelling of nanomaterials in natural waters. *Water Res.* 80, 200–208 (2015).
45. Markus, A. A., Parsons, J. R., Roex, E. W. M., de Voogt, P. & Laane, R. W. P. M. Modelling the transport of engineered metallic nanoparticles in the river Rhine. *Water Res.* 91, 214–224 (2016).
46. Cornelis, G. Fate descriptors for engineered nanoparticles: the good, the bad, and the ugly. *Environ. Sci.: Nano* 2, 19 (2015).
47. Dale, A. L., Lowry, G. V. & Casman, E. A. Much ado about α: reframing the debate over appropriate fate descriptors in nanoparticle environmental risk modeling. *Environ. Sci.: Nano* 2, 27 (2015).

48. Dwivedi, A. D. *et al.* Fate of engineered nanoparticles: Implications in the environment. *Coord. Chem. Rev.* 287, 64–78 (2015).

49. Adam, V. *et al.* Aggregation behaviour of TiO2 nanoparticles in natural river water. *J. Nanopart. Res.* 18, 1–11 (2016).

50. Kaegi, R. *et al. Transformation of AgCl nanoparticles in a sewer system — a field study.* 535, 20–27 (2015).

51. Peijnenburg, W., Praetorius, A., Scott-Fordsmand, J. & Cornelis, G. Fate assessment of engineered nanoparticles in solids dominated media – Current insights and the way forward. *Environmental Pollution*, 1365–1369 (2016).

52. Zhang, Z. *et al.* Transport of cerium oxide nanoparticles in saturated silica media: influences of operational parameters and aqueous chemical conditions. 6, 34135 (2016).

53. Gottschalk, F., Nowack, B., Lassen, C., Kjølholt, J. & Christensen, F. Nanomaterials in the danish environment. Modelling exposure of the danish environment to selected nanomaterials. (Environmental project no. 1639, 2015 from the danish environmental protection agency. 143 pp, http://www2.Mst.Dk/udgiv/publications/2015/01/978-87-93283-60-2.Pdf, 2015).

54. Gottschalk, F., Ort, C., Scholz, R. W. & Nowack, B. Engineered nanomaterials in rivers – exposure scenarios for Switzerland at high spatial and temporal resolution. *Environ. Pollut.* 159, 3439–3445 (2011).

55. Gottschalk, F., Sonderer, T., Scholz, R. W. & Nowack, B. Possibilities and Limitations of Modeling Environmental Exposure to Engineered Nanomaterials by Probabilistic Material Flow Analysis. *Environ. Toxicol. Chem.* 29, 1036–1048 (2010).

56. Sun, T. Y., Gottschalk, F., Hungerbühler, K. & Nowack, B. Comprehensive probabilistic modelling of environmental emissions of engineered nanomaterials. *Environ. Pollut.* 185, 69–76 (2014).

57. Wang, Y., Kalinina, A., Sun, T. Y. & Nowack, B. Probabilistic modeling of the flows and environmental risks of nano-silica. *Sci. Total Environ.* 545–546, 67–76 (2016).

58. Walser, T. & Gottschalk, F. Stochastic fate analysis of engineered nanoparticles in incineration plants. *Journal of Cleaner Production.* 80, 241–251 (2014).

59. Sun, T. Y. *et al.* Probabilistic modelling of engineered

nanomaterial emissions to the environment: a spatio-temporal approach. *Environmental Science-Nano* 2, 340–351, https://doi.org/10.1039/c5en00004a (2015).

60. Song, R., Qin, Y., Suh, S. & Keller, A. A. Dynamic Model for the Stocks and Release Flows of Engineered Nanomaterials. *Environ. Sci. Technol*. https://doi.org/10.1021/acs.est.7b01907 (2017).
61. Keller, A. A. & Lazareva, A. Predicted releases of engineered nanomaterials: From global to regional to local. *Environ. Sci. Tech. Letters* 1, 65–70 (2013).
62. Keller, A. A., McFerran, S., Lazareva, A. & Suh, S. Global life cycle releases of engineered nanomaterials. *J. Nanopart. Res*. 15, https://doi.org/10.1007/s11051-013-1692-4 (2013).
63. Keller, A. A., Vosti, W., Wang, H. & Lazareva, A. Release of engineered nanomaterials from personal care products throughout their life cycle. *J. Nanopart. Res*. 16, https://doi.org/10.1007/s11051-014-2489-9 (2014).
64. Dumont, E., Johnson, A. C., Keller, V. D. & Williams, R. J. Nano silver and nano zinc-oxide in surface waters – Exposure estimation for Europe at high spatial and temporal resolution. *Environ. Pollut.* 196, 341–349 (2015).
65. Gondikas, A. P. *et al*. Release of TiO2 Nanoparticles from Sunscreens into Surface Waters: A One-Year Survey at the Old Danube Recreational Lake. *Environ. Sci. Technol.* 48, 5415–5422, https://doi.org/10.1021/es405596y (2014).
66. Piccinno, F., Gottschalk, F., Seeger, S. & Nowack, B. Industrial Production Quantities and Uses of Ten Engineered Nanomaterials in Europe and the World. *J. Nanopart. Res.* 14, 1109 (2012).
67. EC. Types and uses of nanomaterials, including safety aspects. (European Commission, Brussels, 2012).
68. Zhang, H. *et al*. Processing Pathway Dependence of Amorphous Silica Nanoparticle Toxicity: Colloidal vs Pyrolytic. *Journal of the American Chemical Society* 134, 15790–15804, https://doi.org/10.1021/ja304907c (2012).
69. Sandberg, W. J. *et al*. Comparison of non-crystalline silica nanoparticles in IL-1β release from macrophages. *Particle and Fibre Toxicology* 9, 1–13 (2012).
70. Irfan, A. *et al*. Assessment of Temporal Dose-Toxicity Relationship of Fumed Silica Nanoparticle in Human Lung A549 Cells by Conventional

Cytotoxicity and 1H-NMR-Based Extracellular Metabonomic Assays. *Toxicological Sciences* 138, 354–364, https://doi.org/10.1093/toxsci/kfu009 (2014).

71. ECHA. Guidance on information requirements and chemical safety assessment Chapter R.16: Environmental Exposure Estimation. (2012).
72. Schüler, D., Buchert, M., Liu, R., Dittrich, S. & Merz, C. Study on Rare Earths and Their Recycling Darmstadt. (2011).
73. Maier, M., Gierig, M. & Wegenke, M. Mobilität und Verhalten von Nanopartikeln in der Umwelt. *Bayerisches Landesamt für Umwelt (LfU). Augsburg*. (2016).
74. Maier, M. & Wegenke, M. Untersuchungen zum Vorkommen von Nanopartikeln in Wasser und Luft - 1. Zwischenbericht zum Forschungsvorhaben. *Bayerisches Landesamt für Umwelt, Augsburg (unpublished)* (2017).
75. Erdakos, G. B., Bhave, P. V., Pouliot, G. A., Simon, H. & Mathur, R. Predicting the Effects of Nanoscale Cerium Additives in Diesel Fuel on Regional-Scale Air Quality. *Environ. Sci. Technol.* 48, 12775–12782 (2014).
76. Kaegi, R. *et al.* Synthetic TiO2 nanoparticle emission from exterior facades into the aquatic environment. *Environ. Pollut.*156, 233–239 (2008).
77. Gottschalk, F., Lassen, C., Kjoelholt, J., Christensen, F. & Nowack, B. Modeling Flows and Concentrations of Nine Engineered Nanomaterials in the Danish Environment. *Int. J. Environ. Res. Public Health* 12, 5581–5602, https://doi.org/10.3390/ijerph120505581 (2015).
78. ECHA. Guidance on information requirements and Chemical Safety Assessment, Chapter R.16: Environmental exposure assessment. (Helsinki, Finland, 2016).
79. R Development Core Team. *R: A language and environment for statistical computing*. (R Foundation for Statistical Computing, 2012).
80. Gottschalk, F., Scholz, R. W. & Nowack, B. Probabilistic material flow modeling for assessing the environmental exposure to compounds: Methodology and an application to engineered nano- TiO2 particles. *Environ. Model. Softw.* 25, 320–332 (2010).
81. Coll, C. *et al.* Probabilistic environmental risk assessment of five nanomaterials (nano-TiO2, nano-Ag, nano-ZnO, CNT, and Fullerenes). *Nanotoxicology* (2015).

82. Holten Luetzhøft, H. C., Hartmann, N. B., Brinch, A., Kjølholt, J. & Baun, A. Environmental effects of engineered nanomaterials. Estimations of Predicted No-Effect Concentrations (PNECs). *Environmental project No. 1787, Danish Environmental Protection Agency*. (2015).

10

Nanoparticle Exposure at Nanotechnology Workplaces: A Review

Thomas AJ Kuhlbusch[1,3], Christof Asbach[1] , Heinz Fissan[1,3], Daniel Göhler[2] and Michael Stintz[2]

[1] Air Quality & Sustainable Nanotechnology, Institute of Energy and Environmental Technology e.V. (IUTA), D-47229 Duisburg, Germany
[2] Research Group Mechanical Process Engineering, Institute of Process Engineering and Environmental Technology, Faculty of Mechanical Engineering, Technische Universität Dresden (TUD), D-01062 Dresden, Germany
[3] Center for Nanointegration Duisburg-Essen (CeNIDE), University Duisburg-Essen, D-47057, Germany

ABSTRACT

Risk, associated with nanomaterial use, is determined by exposure and hazard potential of these materials. Both topics cannot be evaluated absolutely independently. Realistic dose concentrations should be tested based on stringent exposure assessments for the corresponding nanomaterial taking into account also the environmental and product matrix. This review

Citation: Thomas AJ KuhlbuschEmail author, Christof Asbach, Heinz Fissan, Daniel Göhler and Michael Stintz, 2011, Nanoparticle exposure at nanotechnology workplaces: A review, Open Access Particle and Fibre Toxicology20118:22, https://doi.org/10.1186/1743-8977-8-22.

focuses on current available information from peer reviewed publications related to airborne nanomaterial exposure. Two approaches to derive realistic exposure values are differentiated and independently presented; those based on workplace measurements and the others based on simulations in laboratories. An assessment of the current available workplace measurement data using a matrix, which is related to nanomaterials and work processes, shows, that data are available on the likelihood of release and possible exposure. Laboratory studies are seen as an important complementary source of information on particle release processes and hence for possible exposure. In both cases, whether workplace measurements or laboratories studies, the issue of background particles is a major problem. From this review, major areas for future activities and focal points are identified.

INTRODUCTION

Research and product developments in the area of nanotechnology have steadily increased especially due to new, beneficial properties of nanomaterials. Nanotechnology as a cross-cutting technology, nowadays used in electrical devices, in construction and composite materials, as catalysts and as antibacterial coatings, is more and more present in workplaces as well as consumer products. This steady increase is accompanied with larger production, handling and processing facilities for nanostructured materials and higher tonnage of nanomaterials.

New nanomaterials, an inherent part of nanotechnological developments, allow on the one hand new products and solutions to e.g. societal problems related to natural resources, drinking water, energy generation and storage, but also raise concerns due to their new specific properties. The major concern is, that the new properties and the high mobility of some nanomaterials may lead to health or environmental effects. This concern has been identified early and was taken seriously by public bodies and the industry. First specific research investigations in toxicology related to particles at the nanoscale were already conducted in the late 1980's [1]. During the last two decades the amount of toxicological research on nanomaterials has increased from less than 10 publications before 1998 to more than 200 in 2010 (ISI Web of Knowledge, 02/2011). A risk, however, may only arise if both a hazard potential of the nanomaterial and exposure exist. Therefore first studies of workplace related exposure were initiated by the International Carbon Black Association (ICBA) in 1998 [2, 3]. In parallel Maynard et al. [4] conducted first studies related to nanomaterial exposure of carbon nanotubes (CNT).

The number of workplace studies and published results has increased significantly since then and a first ISO-guideline on inhalation exposure characterization and assessment has been set-up [5].

Different approaches can be pursued to derive exposure relevant information in workplaces: (a) Studies based on real workplaces and (b) process based studies in simulated workplaces and of simulated work processes. The major advantage of the prior approach is that data from real work conditions are obtained. Still, due to various background aerosols originating from the general work environment or the process itself, extensive and time consuming measurement campaigns have to be conducted. The latter approach, based on simulations in laboratories allows the clear differentiation of a release from the investigated process from background aerosols coming from other sources than the process. It has to be noted that background aerosols may stem from areas outside of the work area and from the process itself. The latter can again be divided into background aerosols released from e.g. electrical motors used in the process [6–8] or unintentionally produced particles from the production process as a side product. This clear differentiation of background aerosols is also necessary to understand what type of background aerosol can be differentiated by which measurement approach.

Process based studies in simulated workplaces and of simulated work processes also enable investigations on how variances in handling, process conditions influence release rates. Release rates are important pieces of information for workplace modelling approaches as investigated and described in detail by Schneider et. al. [9]. Still, laboratory simulations do not represent the real work conditions but represent an important complementary way to derive information on possible nanomaterial exposure at workplaces.

The purpose of this manuscript is to present an overview of publications related to the exposure of engineered nanomaterials in the workplace and processes which may lead to such exposure. Workplace investigations and laboratory simulations are presented and discussed separately since strategies and methodologies employed differ significantly for both approaches. In some cases references to investigations are given in both parts. Those are valuable studies linking process studies with real work conditions, even though this quite often only refers to small bench workplaces.

METHODS, DEVICES AND MEASUREMENT STRATEGIES FOR AIRBORNE NANOOBJECTS AND NANOMATERIALS

Methods and Devices

Measurement and sampling devices for airborne nanoscale particles used in workplace exposure studies can generally be divided into four types, i.e. any combination of size resolved and size integrated with time resolved and time integrated as shown in Table 1.

Table 1: Selected measurement and sampling devices for airborne particles

Size resolved		
Time resolved		
Scanning mobility particle sizer (SMPS)	2.5 nm - 1000 nm > 30 s number size distribution, based on electrical mobility diameter	[10, 11]
Electrometer based mobility particle sizer: fast mobility particle sizer (FMPS)/ engine exhaust particle sizer (EEPS)	5.6 nm - 560 nm 1 s/0.1 s number size distribution, based on electrical mobility diameter	[12, 13]
Electrical low pressure impactor (ELPI)	6 nm - 10 μm 0.1 s number size distribution, based on aerodynamic dia.	[14]
Optical particle sizer (OPS):- laser aerosol spectrometer (LAS)	(> 60 nm) > 300 nm - 20 μm 1 s number size distribution, based on light scattering equivalent dia.	[15]
Inertial spectrometer/time of flight instruments: - aerodynamic particle sizer (APS)	500 nm - 20 μm 1 s number size distribution of the aerodynamic dia.	[16]
Size resolved		
Time integrated		
Low pressure cascade impactor	> 20 nm n.a. mass size distribution, chemical analysis, morphology	[17, 18]
Micro orifice uniform deposit impactor (Moudi)	10 nm -20 μm n.a. mass size distribution, chemical analysis, morphology	[18]
Wide range aerosol system (WRAS)	5.5 nm - 32 μm 5 min number size distribution	[19]
Thermal precipitator (TP)	20 - ca. 300 nm n.a. size distribution	[20]
Size integrated		
Time resolved		
Condensation particle counter (CPC)	5.5 nm - 9 μm 1 s particle number concentration (NC)	[21, 22]
Surface area monitors (e.g. electrical aerosol detector (EAD), nanoparticle surface area monitor (NSAM), LQ1-DC)	10 nm - > 1 μm 1 s aerosol length (EAD), active surface area (LQ1-DC), lung deposited surface area (NSAM)	[23–25]

Aerosol photometer	250 nm - 20 μm 1 s mass concentration	
Size integrated		
Time integrated		
Electrostatic precipitator (ESP)	> 20 nm n.a. chemical analysis, morphology	[26]
Thermal precipitator (TP)	> 20 nm n.a. chemical analysis, morphology	[20]
Filtration (e.g. PM10, PM2.5)	mass concentration, chemical composition	

Size Resolved, Time Resolved

These devices usually use detection principles based on the particles' optical properties or electrical mobility. Hence the results are based on optically or electrically equivalent spheres. In electrical mobility analysis, the equivalence is only valid for the given particle orientation in the classifier, which in case of non-spherical particles may be unknown and hence bias the measurement accuracy. Handling and calibration of mobility particle sizers are standardised in ISO 15900:2009 [27]. Problems of equivalency for different particles also exist for optical detection methods if optical properties vary significantly, e.g. carbon black versus TiO_2. It also is important to take the time resolution into account. Particle size distributions may vary in time scales of seconds which is not resolved by the SMPS (3-5 minutes time resolution) but often used for such measurements. Hence fast particle sizers like FMPS or ELPI should be employed for example to analyse fast changes in the particle size distributions.

Size Integrated, Time Resolved

The particle size weighting is dependent on the metric. Number based detectors give the same signal weight to each particle, while geometric surface area based devices increase the signal weight according to the particle diameter squared. Condensation particle counters (CPC) detect particles from a few nanometres to a few micrometres [22], depending on the model used. Contribution of particles to the surface area is often below detection limit at particle sizes below approximately 20 nm due to the metric immanent weighting of particle size distributions towards larger particles. Diffusion charger based surface area monitors are most accurate in a size range from approximately 20 nm to 400 nm [28]. The lower detection limit is usually not critical due to the low surface area contribution of such small particles in most real particle size distributions. The upper limit, however, may cause more significant errors, because even a small number of these large particles can have a significant contribution to the total surface area.

Size Resolved, Time Integrated

Stahlmecke et al. [29] showed that particles tend to further deagglomerate with increasing pressure difference across an orifice. Hence devices like the low pressure impactor may lead to significantly artificial change of the particle size distribution in the measuring device. A second major error using time integration is the change in physical and chemical characteristics during sampling on the substrate due to interaction with other particles or the gas. One example is the adsorption of volatile organic gases onto collected particles and the substrate.

Size Integrated, Time Integrated

The ESP or TP as shown in Table 1 can efficiently sample down to 20 nm and below. While both types of samplers can generally also sample micrometre sized particles, they can easily be lost in the sampling lines or inlet system. Thermophoretic deposition is more or less independent of particle size up to approximately 400 nm. Hence TPs can provide a homogenous deposit of particles which can also be evaluated for the particle size distribution [20]. The sampling efficiency for fibrous particles in TPs has to be investigated since inefficiencies in collection may exist.

Common measurement methods of time integrated samples applied after collection are the bulk chemical analysis of a filtration or cascade impactor sample, morphological studies using electron microscopy (SEM, TEM) and single particle chemical analysis using Energy Dispersive X-ray analysis (EDX).

More detailed information on measurement techniques for airborne nanoscale particles are given in [15, 30–32].

Despite all the progress made during the recent years with regard to measurement techniques and strategies for exposure assessments towards airborne nanomaterial, there is still the lack of clear performance criteria to achieve comparability and hence nanomaterial specific workplace safety criteria. The development of easy to use devices for the measurements by hygienists is urgently needed even though no agreement on the best metric exists. First personal samplers for time resolved nanoparticle measurements have been recently developed. Such devices can currently not be employed in the breathing zones of workers. In summary, we currently lack information on personal exposure and rely on estimates based on static measurements and first measurements with recently developed devices.

Measurement Strategies

Measurement techniques and measurement strategies have to be optimally combined to allow sensitive and cost effective determination of airborne engineered nanomaterials at workplaces. Measurement strategies can mean tiered approaches as has been suggested by [33] and [34]. Tiered approaches may facilitate cost effective screening of many workplaces if sensitive sensors exists. The NEAT strategy, as developed by NIOSH [33], applies a handheld device which determines particle number concentrations. Particle number concentrations are quite sensitive for nanoscale particles and thus a real-time screening allows identification of possible "hot spots". If the initial screening reveals that the workplace is "clean", i.e. measured concentrations are below a certain threshold value, no further investigations may be necessary. If a nanomaterial release is suspected due to increased concentration levels, the next step of investigations of the tiered approach will be more detailed. This can be personal exposure related, process related or approaches closely linked to toxicological and epidemiological questions.

Directly after the screening detailed measurements related to the above mentioned aims and approaches may be pursued in the areas, which are possibly affected by airborne nanomaterial. Personal exposure approaches will either be based on personal devices and samplers or areal measurements combined with the recording of personal activity patterns to allow the calculation of personal exposure. So far no legal binding framework concerning nanoparticle specific limit values exists. Personal exposure studies related to nanomaterial exposure are therefore dominantly linked to research studies. A good discussion can be found in Maynard and Aitken [35].

Process related approaches will mainly be based on real measurements, allowing for the concurrent use of several and more sensitive devices. Measurement locations will be often close to the process, handling or work activity of interest.

Approaches with the aim of deriving detailed information for toxicological and/or epidemiological studies will use a variety of sampling and measurement devices with possibly health relevant metrics. The variety of devices used for this purpose is high as the metric which best fits to nanomaterial related health effects is still under discussion.

One considerable problem is the distinction of nanomaterial from the background aerosol. Four basically different approaches for background distinction can be differentiated as followed:

- time series approach,
- spatial approach,
- approach based on comparative studies with and without nanomaterial,
- (size resolved) chemical and/or morphological analysis.

The different background distinction approaches are part of the measurement strategies presented above. The task to be tackled determines the combination of needed measurement strategy and devices. Time series analysis is generally coupled with online detection methods. This analysis basically assumes that the concentration determined during no work activity is the background concentration and any increased concentrations during the work activity can be attributed to the process, the nanomaterial or both.

The spatial analysis assumes that a background measurement location is representative for the background at the workplace of interest. Any difference between the determined background and workplace concentrations can be linked to the work activity and the nanomaterial investigated.

Most of the studies published so far use a combined approach of time series and spatial analysis, or link the spatial analysis with morphological or chemical analysis (see Table 2). Morphological analyses are often included when single particle chemical analyses are conducted (SEM-EDX) or distinct features of the nanomaterial can be used for its identification like a fibre structure.

Table 2: Measurement Strategies for background distinction (only articles with original measurements)

	With/without activity	**With/without nanomaterial**
Time series analysis	[36–41]	[42] [2,3]
Time series and, or only spatial analysis	[2][2], [3][2], [33][2,3], [36][1], [38][1], [43–52][2], [53][2,3], [54][2,3], [55][2,3]	

[1]: Only in some cases

[2]: Additional background distinction by chemical analysis (filtration sample or single particle analysis)

[3]: Additional TEM or SEM analysis

[4]: Nearly clean room conditions: [4], not in table

Two approaches are represented by only one published study each (see Table 2). The approach based on comparative studies with and without nanomaterial was published only once [35] but according to the authors' personal communications is often used in industry. This approach assumes any handling and processing of the composite matrix without nanomaterial to be the point of comparison and any changes during handling or processing of the composite with nanomaterial can be attributed to the nanomaterial.

The "clean room approach" was not set up for the assessment of possible worker exposure but needed to avoid product material contamination by ambient dust. Nevertheless, the case in a facility producing carbon nanofibres which flushed a tent with nearly particle free air is an excellent example to avoid background particles from the outside. Still by-product particles either from the production process itself or from other sources such as electrical devices may influence the measurements.

Overall 25 studies related directly to workplace exposure were identified in the literature. More than 50% of these pursued the combined approach based on time series and spatial analysis. Quite often this approach was backed up by chemical and/or morphological analyses since none of the online devices for the detection of airborne nanomaterials is sufficiently selective to unambiguously determine product nanoobjects or their agglomerates. The combined use of the measurement devices significantly enhances possibilities for the distinction of nanoobjects from the background.

EXPOSURE RELATED WORKPLACE MEASUREMENTS

For this review, we have taken into account exposure related nanoparticle measurements at workplaces, where engineered nanoobjects and their agglomerates or products using nanomaterials are processed or used. These include industrial production facilities, processing plants, pilot plant investigations, crafting of nanomaterials (drilling, sawing etc.) as well as research related work area settings. Tables 3 and 4 summarize the reviewed articles. The work activities investigated can be structured along the production pathway: production, handling and refinement of the raw material, bagging and shipping of the nanomaterial, processing of the nanomaterial and work processes with the nanomaterial product. No strict separation according to types of nanomaterials (suspension or powders) is pursued but are indicated by the process investigated, e.g. production via the liquid phase.

Table 3: List of workplace processes and nanomaterials being investigated for possible exposure

Nanomaterial	Production	Handling & refinement	Bagging & shipping	Processing	
				Powder or suspension	In a fixed matrix
Carbon Black	[3]	[3, 37]	[2, 37]	[38]	
CNT, CNF, fullerenes	[4, 33, 46, 48–51, 54]	[38, 33, 4, 46, 45, 50, 49, 54]	[4, 45]	[38, 45, 50–52]	[42, 52]
Ag	[41, 55]	[41]		[33, 53]	
TiO_2	[39, 56, 55]	[39, 33, 56]			
SiO_2	[39, 43]	[39, 43]	[36]	[43]	
Al_2O_3	[39]	[39]		[44, 53, 54]	[44, 54]
Metals	[39, 33, 51]	[39, 33]		[33]	[33]
Metal oxides	[39, 43, 33, 45]	[39, 43, 33, 45]	[45]	[43, 33, 45, 47]	
Others	[39]	[39]		[40]	[40]

Table 4: Summary of the workplace related exposure studies

	Workplace	Type of activity	Nanomaterial	Metric	Results - remarks
[2]	Industrial production	Bagging areas of three plants	Carbon black	PSD(15-675 nm); MC; NC; CC	No significant release of nanoparticles detected, release of agglomerates (> 400 nm) of nanoparticles in all cases of bagging detected if open systems were used; Other sources also significantly influence nanoscale particle concentrations
[3]	Industrial production	Production and pelletizer areas of three plants	Carbon black	PSD(15-675 nm);MC;NC; CC	Significant release of nanoparticles (> 10^6 **#/cm³) and their agglomerates detected in case of a leak in the pelletizing area; in case of good maintenance no significant release of NP from closed production and pelletizing processes; other sources significantly influenced particle number concentrations**
[36]	Toner and printing inks industry	Bag emptying of powders	Fumed silica	NC (< 1 **µm**); PSD (< 1 **µm**); ASA (< 1 **µm**); morph.; CC	Significantly increased[1] **NC (> 100 nm) and ASA detected during bag emptying; confirmed by TEM analysis**
[37]	Industrial manufacturing plant	Manual packaging, warehouse, pelletizing	carbon black	PSD (< 1 **µm**); LDSA (< 1 **µm**)	Higher NC and LDSA concentrations during activity than during non-activity

[41]	Industrial manufacturing facility	Liquid phase process, drying, grinding, handling,	Silver	PSD (15 nm-675 nm); morph	Significant release of particles < 100 nm as well as of agglomerates was observed during all processing steps as soon as the reactor, dryer and grinder were opened, leading to possible exposure even for wet production processes
[45]	Industrial production	Metalloxide production (gas burner) and embedding into a porous oxide matrix, bagging, handling, cleaning and maintenance	MeO (no further information)	NC (10-1000 nm) PSD (14-760 nm), MC PM1 (0.1-1000 nm).	Long term study on possible release of nanomaterial; Significant release of nanomaterial by 'open' production line, handling and cleaning < 1000 nm; Increased NC < 100 nm concurrent with production activity.
[46]	Small commercial nanotechnology production facility	Production of fullerenes (arc reaction), sweeping, vacuum cleaning	Fullerenes	PSD (14 nm - 673 nm), PM2.5 MC, PAH MC	Slightly elevated NC in work area compared to background at one day out of 4 possibly related to cleaning of fume hood; Very good containment of the nanomaterial in the fume hood (production and handling area)
[47]	Industrial production	Wet mill	Lithium titanate metal oxides	NC (10-1000 nm), PSD (300 nm - 10 **μm**) MC (respirable fraction), CC, morph.	Only large agglomerates have been detected
[48]	Industrial production	Bagging and agitation including use of vacuum cleaner during these work steps	Fullerenes	PDS (15 nm-10 **μm**), morph	Release of particles < 100 nm were observed during bagging and vacuum cleaning; also release of particles > 2 **μm** was observed during all work steps, including agitation.

[50]	Industrial production	Production and processing (bagging, handling CNF in dryer, thermal treatment, removal from dryer)	Carbon nanofibers	NC; MC respirable; ASA; photoelectric response; CO and CO2	Elevated NC and MC indicate release of significant amounts of nanoscale particles and their agglomerates; no definite indication on release of single and agglomerated carbon nanofibres.
[51]	Industrial production pilot plant, Industry processing	Production and maintenance (silicon), extrusion of CNT nanocomposites	Silicon; CNT	PSD (5-600 nm); NC; ASA	No changes in PSD and NC was observed during production, but spikes during cleaning of mostly agglomerated silicon (> 200 nm); High NC concentration observed in the extrusion area, but no specific CNT detection method was employed;
[56]	Industrial manufacturing	Production, filtration, bagging	TiO_2, Al_2O_3	PSD (5-600 nm); MC PM1; CC, morph	Wet and combustion production processes were compared and no significant release of particles < 100 nm observed; in one case a bag was overfilled and release of agglomerates > 400 nm observed
[55, 61]	Simulated industry workplace	Compounding of nanocomposites with nanoscale alumina	Al_2O_3	PSD (5.6-560 nm), morph.	Significant release, confirmed by STEM analysis
[4]	Laboratory and industrial production facility	Normal activities during batchwise production of SWCNT: collection, removal, cleaning, opening container, vacuum cleaning	SWCNT	NC (10-1000 nm), MC (size fraction not indicated), morph, CC	Likeliness of CNT exposure during production given; period of exposure relatively short (ca. 1 h) but concentration are sometimes high the exposure nearly pure nanomaterial.
[33]	Laboratory to industrial workplace	Synthesis of nanoobjects, handling and production of composite materials	CNT, CNF, Carbon Nanopearls, fullerenes, TiO_2, Ag, Mn, Co-oxide, Fe-oxide, Al, SiFe, QDs	NC (15-1000 nm) for screening, PSD (300-1000 nm), MC, CC (not size selective)	Increased NC in all three investigated size classes (10-1000 nm, 300-500 nm, 500-1000 nm) indicate Release of nanomaterial during various of the investigated sites; no systematic analysis of the results is presented

[38]	Research Laboratory for use of carbon based ENMs	Transfer of CNMs; sonication in environmentally relevant matrices	Fullerenes, MWCNT; carbon black	PSD (300 - 10,000 nm); NC (10 - 1,000 nm)	Each activity resulted in increased particle number concentrations; TEM images clearly show CNM
[39]	Research laboratories	Scalable flame spray pyrolysis	NaCl, $BiPO_4$, $CaSO_4$, Bi_2O_3, TiO_2, SiO_2, WO_3, Cu/ZN, Cu/SiO_2, Cu/ZrO_2, Ta_2O_5/SiO_2, Pt/Ba/Al_2O_3	PSD (15-675 nm); NC (> 7 nm; > 10 nm), MC (< 1 **µm**; < 10 **µm**)	Concentration in near field and far field higher than in background in 40% of measured cases
[40]	Research laboratories	Plasma enhanced CVD; PVD; compounding of polymers with nanofillers	Nanofillers (not further specified)	PSD (5 nm - 20 **µm**); NC (< 370 nm)	Increased concentrations detected, but likely not caused by ENP release
[42]	Laboratory scale production	Machining/cutting	CNT hybrid composites	NC, PSD (5 nm - 20 **µm**), morph, PM10 MC	Small increases in NC during wet cutting, significant increases (ca. 300,000 #/cm^3) during dry cutting; fibres detected in concentrations of 1-4 fibres/cm^3 during dry cutting.
[43]	Various[2]	Mixing of powder and liquid; filling/emptying oven; suspension spraying; flame spraying	TnO, ZnO, InZnO, SiO_2	PSD (14 nm-20 **µm**), NC (< 1 **µm**), MC (respirable and inhalable)	No evidence of release of ZnO and InZnO during handling; very high concentrations during spraying of silane and flame spraying of SiO2 suspension
[49]	Laboratory scale production	Production by chemical vapour deposition (CVD)	SWCNT, MWCNT	PSD (5 nm - 20 **µm**); morph	SWCNT and MWCNT release was determined in the production area in the fume hood, depending on process conditions; No significant amounts of CNT were detected in the breathing zone of a worker and the background.

[52]	Laboratory scale production and handling	Weighing, mixing with solvent, cutting	raw CNF and CNF composite	NC (10-1000 nm); PSD (10 nm - 10 μm); ASA; morph	Slight increases in NC for weighing, mixing of CNF and wet cutting. TEM picture reveal the release of CNF during these processes. Minor airborne CNF concentration during normal handling; Main increase in PSD for sizes < 400 nm.
[53]	Laboratory handling	Handling in fume hoods of nanomaterial powders, pouring, transferring	Al_2O_3, Silver	PSD (5-600 nm); morph	increased NC in the breathing zone of a worker mainly in size range > 100 nm but also partially < 100 nm during handling activity;
[54]	Laboratory scale production and handling	Growth, removal, shaving and transfer of CVD derived CNT	CNT	PSD (5-600 nm); NC (10-1000 nm); TP, ESP; MC	Neither TEM nor NC analysis reveal a release of CNT during these processes
[62]	Laboratory and industrial production	Four production facilities (2 × TiO_2 by combustion, Ag by plasma and in liquid via citrate), collection of powders in fume hood and in liquid	TiO_2, Silver	PSD (15-710 nm); MC; CC, morph	Lowest number concentration detection for in liquid production, higher particle number concentrations during combustion but also release from electro engines and other side activities

Production

Nanoobjects are either produced top-down by milling and grinding of bulk material or bottom-up starting from nucleation with subsequent particle growth by condensation and/or coagulation. The bottom-up process is the most common industrial synthesis route for nanoobject production. Two major process parameters significantly influence the possible release of nanomaterial: production via the gas [e.g. [51]] or liquid phase [41] and production in an open [45] or closed process [2, 3]. Gas phase production process can further be differentiated into nucleation-condensation (metals), vapour deposition (CNTs) and gas-to-particle conversion via oxidation (metal oxides). The latter differentiation itself does not influence the possibility of release but may determine whether the process is open or closed (references given after the brief description of the process indicate corresponding exposure related studies).

Handling and Refinement

Once the nanomaterial is produced, it is removed from the process by filtration or opening of a reactor. Relevant, sometimes manual work steps are filtration [3], pelletizing [2], cleaning of raw nanomaterial [56], drying [41], grinding and milling [56].

Bagging and Shipping

The next working steps are bagging and shipment of the nanomaterial. The nanomaterial may either be handled as liquid suspension or dry powders. The latter is often discussed and investigated as a source for airborne exposure [2, 4] while the prior significantly decreases the likeliness of airborne release. No investigations related to bagging/filling and shipping of suspensions have been published so far.

Processing

The area of activities related to processing of nanomaterials covers a wide range such as mixing nanomaterial powders in liquids, cutting, drilling etc. of composites, drying and spraying. Therefore one single study can rarely cover all processing activities of one nanomaterial. E.g. the handling and mixing of CNF was investigated by Mazzuckelli et al. [52] but similar processes for TiO_2 or other nanomaterials are not published. We may differentiate processing steps with the nanomaterial itself as a powder or in a suspension and those processes, e.g. drilling or cutting, when the nanomaterial is embedded in a matrix.

Work Process with Nanomaterial Products

The use of nanomaterial coated products at medical workplaces, workplaces using nanomaterial products, such as dry cleaners, parquet sanding are also nanomaterial related workplaces. Exposure scenarios towards the nanomaterial for these workplaces have so far not been published to our knowledge and are not further covered in this review.

An overview on which process was investigated including which materials is given in Table 3. It can be seen that most of the materials and processes have been investigated. Still quite a few processes were only part in overview studies, e.g. by Methner et al. [33] or Möhlmann et al. [43], which summarize results but do not go into detail and give no clear results whether the actual nanomaterial was released or not.

The lack of a harmonized approach concerning measurement strategies and techniques, metrics and size ranges as well as the data analysis procedures complicates the summary of the studies listed in Tables 3 and 4. The different measurement strategies and measurement devices were introduced in the prior section. From Table 4 it is evident that all studies used particle number concentrations (NC), either directly determined by a CPC or derived from particle size distribution (PSD) measurements, in their analysis for possible exposure. This is not surprising since number concentration measurements are relatively easy, cheap and very sensitive towards nanoparticles. The main problem for a combined assessment of health effect studies related to NC and PSD is that no defined lower detection limits for particle size are used. This leads to difficulties in the comparison of absolute values, especially in case of number concentrations, where contributions of sub-10 nm particles can be significant.

21 out of the 25 studies summarized here also determined the particle size distribution down to below 100 nm particle size. This is seen to be essential to actually allow for the differentiation of larger agglomerates from the more mobile particle size fraction below 100 nm. While the time resolution of NC measurements is normally relatively high (1 Hz), time resolution of PSD devices may become crucial when studying processes with quickly changing aerosols. It is also important to note that determined size distributions, either by electrical mobility or optical properties are based on diameters of electrically or optically equivalent spheres, which may not be very meaningful for non-spherical particles.

Slightly more than half of the studies determined also the mass concentration (MC). Various devices, e.g. optical particle sizer based measurements, stationary and personal filtration samplers, and PM1, PM2.5, PM10 samplers were employed. The use of filtration samplers and MC determination is seen to be important to enable the linking with conventional workplace exposure assessment and legislative limits. Nevertheless, this metric is dependent on the particle volume, i.e. particle diameter to the third power and is hence very insensitive to nanomaterials in the lower submicrometer size range.

The importance of the use of electron microscopy (EM) for nanomaterial exposure related studies is stressed by the fact that 12 studies included this time consuming analysis. Even though the use of EM may not be seen as quantitative it is still the only method, which allows for a definitive identification of product nanomaterial, especially when linked with single particle chemical analysis. Good examples are here the detection of CNTs and CNFs.

Eight studies also used consecutive chemical analysis to derive more information on whether the airborne material may have been the nanomaterial in use. Bulk chemical analysis can be used for the different size fractions as indicators, but only single particle analysis can provide definitive differentiation from the background.

The possible importance of the particle surface as exposure metric was presented by Oberdörster et al. [57] and stressed by several other researchers [58, 59]. Easy to use, portable devices became available only in recent years an4 studies already used these devices. None of the currently available measurement devices measures the geometric surface area of particles, but the lung deposited surface area or "active" surface area. Although the delivered metric is not the geometric surface area of the particles, the devices output can directly be linked to health relevant toxicological reaction mechanisms. The practicability is a further advantage of this measurement technique. On the other hand, this metric is increasingly insensitive for decreasing particle sizes.

The lack of any uncertainty and detection limit discussion in all of the studies is another problem in the assessment of the published exposure related studies at workplaces. Kuhlbusch et al. [56] have recently published a first uncertainty estimate for the method they employed. They showed that no single uncertainty value can be given due to the high dependency on local conditions, like variance in background concentrations. If two or more devices are used simultaneously to obtain a spatial distribution, the uncertainty is furthermore influenced by the measurement accuracy and intercomparability of the measurement devices. Even two devices of the same type can show deviations up to approximately 30% [60, 44].

Some general conclusions on the focal point, possible airborne exposure to nanomaterials, can be drawn from the studies. 22 of the 25 studies indicate release of nanomaterial particles > 100 nm. The remaining three studies reported increased values in the work area but stated that the elevated concentration may not be attributed to the nanomaterial. Two of those were laboratory scale production studies related to CNTs. This ambiguity in the identification is much more common for the reported values for particles below 100 nm particle size. 13 of the 25 summarized workplace studies in Table 4 indicate release of nanomaterial below 100 nm particle size. But, only few studies were able to clearly identify the nanomaterial, Kuhlbusch et al.

[3], Fujitani et al. [48], Tsai et al. [49]. In other cases, results were much less clear as in Yeganeh et al. [46] and Evans et al. [50]. This listing shows that the identification of nanomaterial exposure is not straight forward and that well described and harmonized methodologies are needed a) to unambiguously link elevated particle concentrations to the nanomaterial under investigation and b) come to quantitative results. The latter is extremely difficult and possibly needs process studies on the laboratory scale.

NANOPARTICLE RELEASE STUDIES UNDER LABORATORY CONDITIONS

Systematic and quantitative analysis of processes and materials can be performed under defined boundary conditions in the laboratory. Therefore, potential sources must be previously identified and selected. Relevant process-related parameters have to be chosen for the transferability to real exposure situations. To meet these requirements, some studies used standardized procedures for the experimental process simulation [53, 63–66]. Due to the multitude of processes which lead to the release of particles, the number of studies on new characterization methods is steadily increasing [29, 54, 66, 67].

Process control (e.g. avoidance of background concentration and particle contamination, sampling), conditioning of the generated aerosol (e.g. dilution, neutralization, size selective particle deposition), and the combination with suitable measurement technology are crucial for the quality of investigations on particle release in the laboratory. Particle number based aerosol measurement devices, originating from clean room monitoring, enable the highest sensitivity for the quantification of low particle amounts. Therefore the nanoparticle release is expressed as a particle number released from a sample of defined size [68].

Nearly all of the reviewed laboratory studies are based on reducing background concentrations. The easiest way to ensure a stable and well known background concentration is to isolate the test room atmosphere from outside influences as performed by Bello et al. [42, 54, 69]. Lower background concentrations require encapsulation and purging with filtered air or gas as realized by the most studies. The lowest background concentrations were achieved by using laminar flow benches which also provide larger working space as employed by Vorbau et al. [64] and Göhler et al. [67].

To ensure the comparability of different tests or to estimate the relevance for a certain workplace situation the volumetric flow over the treated sample, the analysed volumetric flow of the measurement device and the dilution ratio must be measured to calculate the number of emitted nanoparticles (defined as nanoparticle release) from the measured particle number concentration and particle size distribution. Furthermore, the comparability requires the reporting of treated sample size, for instance sample mass. Relating the released nanoparticle number to this sample mass delivers a kind of emission factor of a certain material in a certain treatment process.

Generally, nanomaterial release studies can be structured in following steps:

- reducing background particle concentration
- sample treatment and/or process simulation
- nanomaterial release and aerosolisation
- aerosol sampling and conditioning
- aerosol/particle characterization

Nanomaterials are subjected to mechanical, thermal and environmental stress situations during production, processing and use. Published studies, based on the characterization of the particle release into air due to individual treatment processes, can be roughly classified by the investigated nanomaterial group (powders, suspensions, coatings, composites) as shown in Table 5. According to these groups the experimentally simulated treatment processes (shown in column two) are different like fluidized bed processes for powders or sanding processes for coatings. Comparable processes are separated additionally. Moreover, the processes are ordered downwards with increasing energy input.

Table 5: Studies focused on the particle release of ENMs by laboratory testing (column 1 indicates the different subcategorisations)

Study	Process	Materials	Instrumentation	Metric	Results
Powders					
[53]	Rotating drum test (free fall, stirring, ...)	TiO_2 (Aeroxide P25), ZnO	SMPS, APS, MOUDI	PSD (15 nm - 20 **μm**), NC, DI	< 10% particles < 100 nm60% particles < 1 **μm**

[65]	Rotating drum test (free fall, stirring, ...)	TiO_2, SiO_2, FeO(OH), $Mg_3Si_4O_{10}(OH)_2$, Al_2O_3	FMPS, APS, Filtration	PSD (5.6 nm - 20 **μm**), NRP (0.5 **μm** - 20 **μm**), NRP (5.6 nm - 560 nm), DI	Undefined fraction of particles < 100 nm
[66]	Rotating drum test (free fall, stirring, ...)	Organoclay, Bentonite	FMPS, APS, Filtration	PSD (5.6 nm- 20 **μm**), NRP (0.5 **μm** - 20 **μm**), NRP (5.6 nm - 560 nm), DI	Undefined fraction of particles < 100 nm
[70]	Free fall	TiO_2 **(G5), SiO_2** (Aerosil 200)	ELPI; SEM (ELPI)	PSD(30 nm - 10 **μm**), MC, MC/M	Fraction of particles < 100 nm
[71]	Fluidized bed	TiO_2 (Aeroxide P25)	SMPS (LDMA, NDMA), APS; TEM (ESP)	PSD (4 nm - 20 **μm**)/NC_{max}	10% particles < 100 nm 70% particles < 1 **μm**
[4, 72]	Vortex shaker (fluidized bed, agitation)	SWCNT, alumina powder	SMPS (LDMA, NDMA), APS	PSD (4 nm - 20 **μm**), NC	Fraction of particles < 100 nm, alumina powder released more NP than SWCNT
[73]	Vortex shaker (fluidized bed)	SWCNT, MW-CNT, TiO_2, ZnO	SMPS, HHCPC, APS, OPC	PSD (10 nm - 20 **μm**), NC (10 nm - > 1 **μm**)/V, NC (10 nm - > 1 **μm**)/M	Fraction of particles < 100 nm
[74]	Shaker method	MWCNT	SMPS, APS, TEM(ESP, CI)	PSD (14 nm - 20 **μm**), NC(dt)	Fraction of particles < 100 nm; peak at 200-300 nm
[75]	Fluidized bed with oscillating sieve plate	MWCNT	SMPS, SEM (TP)	PSD (< 1 **μm**), NC(dt)	Fraction of particles < 100 nm
[29]	Stirring and dispersing in orifice (leak in pressurized vessel)	TiO_2, CeO_2, $SrCO_3$, TiZrAlO	SMPS	PSD (14 nm - 736 nm), fractions, relative values	Increase of the fraction of ENPs by increase of the overpressure (up to 12%)
Suspensions					

[38]	Weighing/ transferring of powders and sonication of suspensions	fullerenes, MW-CNT, CB	HHCPC, HHPC, TEM-EDX (filtration)	PSD (0,3 **μm** - 10 **μm**), NC (10 nm - 1 **μm**)	Suspension sonication leads to droplets with embedded ENM
[77]	Spraying	suspensions with and without Ag	SMPS, TEM(ESP)	PSD (10 nm - 500 nm), NC (< 100 nm), NC(< 500 nm)	High fraction of particles < 100 nm
Coatings					
[63]	Weak abrasion process (Taber Abraser)	PVC layer with/ without ENPs (nanoclay)	SMPS, CPC	PSD (5 nm-1 **μm**)	
[64]	Weak abrasion process (Taber Abraser)	coatings with/ without ENPs (ZnO)	SMPS, CPC; SEM/TEM-EDX (ESP)	PSD (16 nm - 626 nm), NC (> 6 nm), wear mass	Very low concentrations, ENPs still embedded
[81]	UV light, wind erosion, scrabing	coatings with TiO_2	SMPS	PSD (15 nm - 661 nm), NC (15 nm - 661 nm)	Comparison with non-doped samples is missing
[54]	Shaving (razor blade)	CNT	FMPS, HHCPC, SEM & TEM - EDX (TP)	NC (10 nm -1 **μm**), PSD (5.6 nm-560 nm)	No significant change in concentration; no free CNTs were observed
[7, 8]	Sanding process (orbital sander)	coatings with/ without ENPs (TiO_2, CB, SiO_2, $CaCO_3$)	FMPS, APS(ESP)	PSD (5.6 nm-20 **μm**)	General release of NP, spark particles contamination
[67]	Sanding process (Dremel)	coatings with/ without ENPs (ZnO, Fe_2O_3)	FMPS, CPC, OPCSEM/TEM-EDX (ESP)	PSD (5.6 nm-20 **μm**), NC (< 100 nm), NRP (< 100 nm), NRP (< 10 **μm**), swarf mass; material, morphology	General release of NP but ENPs still embedded in the matrix
Composites					
[82]	UV-light; weak abrasion process (Taber Abraser); customized sanding	POM with/without CNT, PA with/ without SiO_2, cement with/ without CNT, cement with/without CSH	SMPS, UNPA, SEM, AUC, XPS, SIMS	PSD (14 nm-820 nm), Morphologie, material	No free SiO_2-particles or CNTs detected

[42]	Dry and wet cutting (band-saw; rotatory cutting wheel)	composites with and without CNT	FMPS, APS, HHCPC, DT, SEM/TEM-EDX (ESP, TP)	PSD (5.6 nm-20 µm), NC (5.6 nm - 560 nm), NC (0.5 µm - 20 µm) MC, material, morphology	No free CNTs observed
[69]	Dry and wet solid core drilling	Composites with and without CNT	FMPS, APS, HHCPC, DT, DC, SEM/TEM-EDX (ESP, TP) WRASS+ICP-MS	PSD (5.6 nm-20 µm), NC, MC (< 35 µm), SA; material; morphology	Smoke generation, free CNT clusters observed

Powders

Based on the various application processes (e.g. transport, dosing, filling) of powders and their characteristic parameters (e.g. energy input, flow velocity, sample amount), different methods have been developed in the past for the investigation of material dependent particle release, the so called dustiness. Additional material dependent parameters, like powder flow properties also affect the results of such dustiness tests and limit their comparability. Nevertheless different types of test setups are described in national standards and two test procedures based on different drop conditions (rotating drum test, continuous drop test) were introduced as international standard (EN 15051:2006). These methods, developed for MC assessments, were taken, modified and combined with aerosol measurement devices by Schneider & Jensen [65], Jensen et al. [66] and Tsai et al. [53] for the characterization of the particle release. Schneider & Jensen [65] and Jensen et al. [66] investigated the particle size distribution and particle number concentration of different metal oxides powders (TiO_2, SiO_2, FeO(OH), $Mg_3Si_4O_{10}(OH)_2$, Al_2O_3), bentonite and organoclay powders. Typical results show a number based submicrometer fraction of about 60% and < 10% for particles < 100 nm. Similar results were obtained by Tsai et al. [53] for TiO_2 and ZnO metal oxide powders with slightly modified measurement devices (SMPS, APS, and MOUDI). Isbaseta & Biscans [70] investigated TiO_2 and SiO_2 in free drop conditions. Despite large deviations in the PSD determined by an ELPI, both powders show a release of nanoparticles of 10 $mg{\cdot}m^{-3}$ an40 $mg{\cdot}m^{-3}$ for TiO_2.and SiO_2, respectively, for the same initial sample mass.

Studies based on fluidized bed processes represent a second test type for dustiness. Maynard [71] aerosolized TiO_2 powder in a two component fluidized bed to determine the agglomerate size distribution between 5 nm and 20 µm. Furthermore, airborne particles were electrostatically-precipitated for subsequent TEM analysis. Results showed about 10% of the particle number concentration to be < 100 nm diameter and 70% in the submicrometer range. Baron et al. [72] and Maynard et al. [4] introduced a new method for agitating CNTs and metal oxide powders to achieve higher particle concentrations. Powder samples or mixtures of powders and beads were filled in a HEPA-filtered air purged shaker-agitated centrifuge tube. The PSD measurements (3 nm - 20 μm) showed that fumed alumina powder released more particles < 100 nm than the investigated SWCNT powder. The agitating principle of Baron et al. [72] and Maynard et al. [4] was also used by Ogura et al. [73] to characterize the amount, PSD and morphology from different nanostructured powders (TiO_2, ZnO) and carbon materials (SWCNT, MWCNT, fullerenes). In addition to the PSD, Ogura et al. [73] relate the measured number concentrations to the applied sample volume or sample mass. Lee et al. [74] used a similar shaker-principle for the PSD characterization of MWCNTs. They also employed TEM-analyses. Results of atomized, dried and neutralized suspensions of the same MWCNT and ultra-pure water show similar peak diameters at approximately 200 nm up to 300 nm in comparison to the shaker-generated aerosol. In each case fractions of particles in the nanometer size range were observed. An extension of the fluidized bed to aerosolize CNTs was presented by Plitzko et al. [75]. To improve the aerosolisation process, the sieve plate of the fluidized bed was agitated by a shaker. First results based on SMPS measurements and TEM investigations on TP-precipitated particles show a considerable release of CNTs with material specific fractions of particles/agglomerates in the nanometer size range. The amount of particles with electrical mobility diameters < 100 nm was determined to be between 10% and 60% of the determined number based particle size distribution < 1000 nm.

The procedures and results presented up to here are based on a low energy input to the powder. High energy processes, on purpose or by accident, can result in significantly higher energy input to the nanomaterial. Therefore, worst-case scenarios have to be investigated. The particle release due to a leak in a pressurized vessel was experimentally simulated by Stahlmecke et al. [29]. Different metal-oxide powders (TiO_2, CeO_2, $SrCO_3$, TiZrAlO) were aerosolized using a pressurized beaker in combination with a magnetic stirrer and passed through an orifice. Results are given as PSD and relative concentrations based on the comparison with reference measurements of powders dispersed without orifice, mimicking the leak. In the measured size range < 736 nm, the amount of particles < 100 nm of the number weighted particle size distribution increases from 1% up to max. 12% with increasing overpressure from 0 kPa to 140 kPa, depending on the material. Another worst case scenario, preferred by NIOSH and introduced in ISO DIS 12025:2010 [68], is the method described by Boundy et al. [76], where a small amount of powder - aroun4 g - is completely sucked and dispersed in an evacuated bottle.

Nearly all powder studies suffer from incomplete determination of the energy input during sample treatment. Some powder drop tests were affected by the material properties, e.g. flowability, themselves. Furthermore repeated treatment of powders in rotating drum or fluidized bed can generate secondary agglomerates. In this case the number of released particles rapidly decreases during the test [72, 75]. Fast measuring devices may help to avoid this problem [65, 67]. Furthermore, treatment processes stressing a well-defined sample only once allow a straight correlation of the measurement data to the treatment process. Repeated energy input leads to permanent changes in the experimental conditions.

Suspensions

Johnson et al. [38] investigated the particle release of fullerenes, MWCNT and carbon black into air due to the sonication of suspensions in comparison to different laboratory handling processes, especially the weighing and transferring of powder material. Size specific particle number concentrations (300 nm - 10 μm) and total particle number concentration (10 nm - 1 μm) were analysed. In addition, the generated aerosols were filtrated for subsequent TEM analysis. Johnson et al. (2010) [38] found that sonication leads to the release of water droplets with embedded nanomaterials. CPC measurements showed an increase in concentration by sonication in comparison to the dry powder bulk handling, while OPC measurements showed a decrease in the concentration of coarse particles. It remains unclear whether all droplets contained nanomaterial.

Spray simulation with nano-silver and nanoparticle-free suspensions were carried out by Hagendorfer et al. [77] inside a particle free glove box. Two different types of commercial spray dispensers (pump spray, propellant gas spray) were investigated. The generated aerosols were dried by a thermodesorber before PSD measurement and sampling by an ESP for subsequent TEM and EDX analysis (0.8 m away from the generation zone). The pump spray experiments showed no significant increase in the number concentration (10 nm to 500 nm) in comparison to the background. This can possibly be explained by the formation of only large droplets which deposited inside the glove box before sampling. Propellant gas spray, on the other hand showed a broad PSD. The silver particle sizes, determined with SMPS and TEM, depended on the droplet size generated by the dispenser. The measured number based PSDs of the propellant gas spray generated aerosol show a fraction of around 80% < 100 nm.

Coatings

Engineered nanoparticles are often intentionally embedded in the matrix material of a coating, which can be understood as a thin layered composite on a substrate. Several studies were performed to determine potential exposure of consumers by typical application processes of such coatings like abrasion, scraping or sanding. Besides the determination of the particle concentration and the PSD, additional appropriate chemical and morphological analyses (e.g. SEM, TEM, EDX, ICP-MS) are necessary for the distinction of unintentionally produced particles in the nanoscale and the engineered nanoobjects in the airborne emissions.

Vorbau et al. [63] employed a commonly used and standardized Taber Abraser (DIN 68861-2:1981 [78], DIN EN 13523-16:2005 [79], DIN EN ISO 7784:2006 [80]) for determining the particle release from different coatings with and without ZnO nanoparticle additives. This device consists of two abrasion wheels, which act on the sample surface. Investigations were performed by operating the Taber Abraser in a particle free environment inside a laminar flow bench. Due to the weak abrasion process, the nanoparticle concentration was too low for determining PSDs by SMPS. The average total particle concentration based on the CPC measurements ranged between 1 cm^{-3} to 20 cm^{-3}. No significant differences were observed between coatings with and without added nanomaterial. TEM images of electrostatically collected wear particles showed the nanomaterial only embedded in the matrix. In another study Guiot et al. [63] used the Taber Abraser test method for investigations on polyvinylchloride layers with and without nanoclays on a PET substrate. In contrast to Vorbau et al. [64] the generated and aerosolized particles were locally sampled near the abrasion zone. The abrasion zone was encapsulated and purged by HEPA-filtered air. Measurement data show an increase in the particle concentration for nanoclay-doped samples. The authors subtracted the measured PSD of the PVC layer without nanoclays from that with nanoclays. This calculation assumes same particle release rates in both experiments for balancing particle amounts, which was not shown. Furthermore, the difference in the released number of nanoparticles may not represent the nanoclay release due to changed mechanical matrix properties introduced by the addition of nanoclay.

Nanoparticle release due to light at different wavelengths, air flow and mechanical scraping for surface coatings containing nanomaterials were investigated by Hsu and Chein [81]. Coatings with different TiO_2 additives on wood, polymer and tile were tested in a closed, purged chamber and PSDs were measured (15 nm - 616 nm). The highest particle concentrations of about 630 cm^{-3} were found for the coating on tile and wood during parallel use of UV-light, airflow and scraping. No comparative measurements of coatings without added nanomaterial were performed, so a clear interpretation of the measured data is not possible. Bello et al. [54] investigated the particle release during the removal of CNT forests, which were artificially grown by CVD on a silicon substrate, by cutting with a razor blade. No significant difference between background concentration and material handling was observed using an FMPS and a CPC. SEM and TEM-images of electrostatically and thermophoretically collected and filtrated aerosol particles showed typical background particles but no CNTs, whether individual or bundles.

A commercial hand-held orbital sander with an internal dust removal fan was operated by Koponen et al. [7, 8] for the characterization of the sanding dust. The exhaust air of the sander was connected to an exposure chamber for determining the PSD (5 nm - 20 µm). Additionally, particles were electrostatically sampled for subsequent physicochemical and toxicological analysis. Coatings containing and not containing engineered nanoparticles were investigated. Some coatings showed higher others lower release rates when engineered nanoparticles were added to the coating. Sanding of surface coatings with and without ZnO and Fe_2O_3 nanoparticle additives was investigated by Göhler et al. [67]. The sanding zone was encapsulated and a particle free environment was provided by a laminar flow bench. PSD (5 nm - 20 µm) and number concentrations (CPC) were measured and ESP samples analysed by SEM, TEM and EDX. Despite a considerable generation of nanoscale particles due to the sanding process, neither matrix-free engineered nanoparticles nor significant differences in the particle size distribution between coatings with and without engineered nanoobjects were observed in both studies.

Composites

Particle release due to Taber Abraser tests, customized sanding and UV-light weathering was compared by Wohlleben et al. [82] for different composites containing SiO_2, CNT, CSH nanomaterial. In each case the morphology, the size distribution and surface chemistry of the wear powders were analysed. Aerosol particles were measured by SMPS for the first two processes. Only the polymer-composite with CNTs behaved significantly differently to the same material without nanoobjects, where increased UV-light adsorption resulted in faster matrix degradation.

The ability for the release of CNTs from composite materials was also investigated by Bello et al. [42, 69] during wet and dry cutting and solid core drilling. In both studies the same types of CNT-composites were analysed. Bello et al. [42] used a band saw for the simulation of a dry cutting process, while a rotating cutting wheel was operated for the wet cutting process. Samples for the analyses were taken at two different locations - near the source for the dry cutting process and at the breathing zone. The PSD (5 nm - 20 μm), NC (10-1000 nm) and MC (PM2.5, optically determined) were measured. Analysis for particle chemistry and morphology were carried out by SEM, TEM coupled with EDX on ESP and TP substrates. A considerable generation of nanoscaled particles during the dry cutting process was observed. Results show no significant difference between samples with and without CNTs with regard to the particle number concentration. SEM and TEM analysis showed no free CNTs and no CNTs which are bundled or attached to coarser particles.

A commercial drill press was employed by Bello et al. [69] to simulate dry and wet solid core drilling. The latter one was simulated by continuous spraying of deionized water to the drilling zone. In comparison to Bello et al. [42] a WRASS was added to the measurement devices in [69] for size selective particle deposition and subsequent ICP-MS analysis. During the drilling processes smoke generation was observed. Also clusters of CNT were determined with TEM. Whether these were released directly from the drilling process or originate from contaminations remains unclear.

DISCUSSION AND CONCLUSIONS

Two different basic approaches to derive information on possible workplace exposure were summarised and the state of the art presented in this review. It is evident that both approaches, workplace measurements and laboratory studies, are needed for a concise assessment to derive e.g. model based predictions on possible release of nanomaterial from a given process.

All studies can be ordered in a systematic scheme of different workplaces, work processes and nanomaterials. This systematic evaluation shows that all basic parameters have been pursued and published in the peer-reviewed literature. Thus, the amount of data becoming now available will allow first meta-analysis investigating material structure and release relationships in detail. Despite the major progress made in recent years, severe open issues still exist hindering the set-up of a coherent and concise exposure assessment for nanomaterials at workplaces. A coherent approach for the likeliness of release and workplace exposure assessment for all relevant nanomaterials and work processes as e.g. outlined in Table 3 is recommended.

Measurement metrics and corresponding measurement technology remain to be an extremely important issue. The metric best related to possible health effect is still not identified. The lack of comparability always became evident, when trying to compare results from the various studies in this review, because a harmonized approach for data evaluation concerning metric, size range etc. is still missing. This lack hinders the development of general conclusions.

Particle number concentrations and particle number size distributions are the most commonly used metrics within the reviewed workplace and laboratory studies. This approach is currently seen as the one to further develop due to the high sensitivity of the metric to airborne nanoobjects as well as due to the availability of the measurement devices. The latter allow the use in tiered approaches starting with a screening e.g. only measuring a size-integrated concentration quantity such as the number concentration. Intensified measurements can be conducted in a second tier in identified areas of increased concentration.

A major drawback of current state of the art measurement devices is their lack of differentiation of background from nanomaterial related particles. Aerosol mass spectrometer is currently the only instrument sizing and chemically analysing nanoscale particles online. Such instrumentation, once it is capable of also analysing metals and metal oxides, which is currently in development, would be an ideal analytical instrument to be used for areal measurements at workplaces if costs and size of the instrument are neglected. On the other hand particle number concentration and particle size distribution measurements at workplaces can at best only be viewed as indicative measurements for the presence or absence of airborne nanomaterials but without a definitive proof. The latter can only be provided by particle sampling with subsequent electron microscopy, if possible linked with a single particle chemical analysis.

A second drawback may be the sensitivity and comparability. While CPCs can reliably measure concentrations down to 0.001 cm^{-3}, size resolved measurements for nanoscaled particles need significantly higher concentrations. In addition, the commonly noisy background makes it impossible to detect a release which only amounts to a small fraction of the total concentration. In case of the need for detection of concentration increases of a few particles per cm^3 other devices and methods have to be employed. An obstacle related to this topic is the current lack of information of lower and upper detection limits related to measurement devices and especially when combined with a measurement strategy. Further uncertainty arises from the calibration of the devices which sometimes may differ by up to 30% [60]. Work on the uncertainties and detection limits is certainly urgently needed to achieve data qualities good enough for comparison of results from different studies.

The method mostly used for the identification of nanomaterials is electron microscopy (SEM or TEM) coupled with single particle chemical analysis such as EDX. This method was regularly employed in different studies but is not standard in routine workplace assessments due to the high demand of person hours. A difficulty for this approach is the limitation of the measurements. Nevertheless the use of single particle analysis, combining morphological and chemical information, is currently the only approach proofing the presence of engineered nanomaterials. A definitive proof of their absence may not be possible due to the limitation of the subsequent quantification to a few thousand particles, even if automated particle identification software is used.

There is still an urgent need for a systematic approach of harmonization and standardization. The needed areas to be covered are test procedures simulating workplace activities and processes as well as coherent workplace exposure assessments. First steps in simulating workplace activities and work processes are investigations on sensitivities of nanomaterial emissions to specific work parameters such as in sanding on the type of sanding paper, the acting force onto the surface, rotational speed. Further steps in exposure assessments is the European, better worldwide, agreement on a harmonized measurement metric, strategy, and data treatment and analysis including statistics.

From the discussion above and the outcomes of the studies presented in this review, we conclude that a tiered approach is viewed as most practical for workplace and laboratory measurements, because complete measurement campaigns are very time and hence cost intensive and may only be necessary if there is evidence of an increased particle concentration. The proposal brought forward by NIOSH [33] and the consent report [34] are scientific reasonable and pragmatic starting points for further refinements. Laboratory studies are the consequence of the tiered approach.

Some studies reported a release of particles. In fewer studies engineered nanoscale particles were observed. The latter was mainly caused by maintenance problems, open gas phase production processes, open handling of nanopowders or smoke generation during processing.

A release of agglomerated nanoobjects, mainly > 300 nm in the number weighted diameter, was regularly observed, especially during open handling of dry nanomaterials [2, 3]. Release of nanomaterials < 100 nm was only observed in the few special cases mentioned above. The use of fume hoods and appropriate ventilation systems seemed to significantly reduce potential exposure concentrations.

The review of the laboratory test procedures shows that many workplace related processes are currently simulated in the laboratory. While some approaches are quite advanced we still lack a coherent, systematic approach over all work related processes as well as studies on single simulations. Questions like: How shall abrasion tests be pursued? How does the normal force of the abraser influence the release? Shall heating of the sample be kept at a minimum? Shall the test simulate worst case scenarios? How to avoid background particles? have to be answered before a simulation method can be performed. This area, important for modelling of work processes and subsequent possible exposure, has to be further developed on the basic research level as well as in view of a standardized method.

Generally all reviewed studies underline that the amount of released nanoparticle is the result of the combination of the treatment process and the employed material. All powder handling processes released some nanoobjects. Release of free engineered nanoparticles was not observed in laboratory studies of workplace related processes based for treatment processes on coatings and composites because the nanomaterial was still embedded in the matrix material.

Still one of the major issues to be tackled in the near future is the question of how to detect and define level of detection needed for workplace safety assessment. In the case of a specific health hazard of a nanomaterial, lower detection limits may be down to single particles are necessary. Generally, particle number concentration based exposure values may be sufficient in most cases. When assessing the metric to be used for limit values, possible other parameters, such as particle surface area and particle reactivity [e.g. [83]], have to be evaluated. The use of mass concentration measurements is currently viewed as being too insensitive to assess toxicological effects related to airborne nanomaterials.

A final point revealed by this review is that a certain set of minimum information is needed for all workplace related studies, either real workplaces or laboratory simulations. Release assessments and comparability between different studies require sample size specific nanoparticle release data, which can only be obtained by simultaneous measurement of several more parameters than simply particle size distributions and number concentrations. Nevertheless, in each case, whether workplace studies or laboratory investigations, comparisons with reference activities, reference materials or different treatment processes are of fundamental importance for the discussion on possible nanomaterial exposure. The nanomaterials and processes studied have to be described carefully and in sufficient detail, including contextual information, to be able to compare the results with those of other studies. All methods employed to identify the corresponding nanomaterial should be described and a clear conclusion given if a release or exposure was determined. A last but very important piece of information is the reporting of particle sizes and size resolved concentrations. Integrated particle number and surface area concentrations alone may be used as indicators in a tiered approach, but do not describe exposure in sufficient detail.

ENDNOTES

[1] The wording and definitions used in this review article is based on ISO/TS 27687 (2008) [84].

ACKNOWLEDGEMENTS

This review was financially supported by the Verband der Chemischen Industrie e.V. (VCI, representing about 1600 chemical companies in Germany), but does not necessarily represent their opinion. Special thanks are given to Dr. Stefan Engel and Dr. Matthias Voetz for their discussions and support during the preparation of the review.

REFERENCES

1. Oberdörster G, Ferin J, Finkelstein J, Wade P, Corson N: "Increased pulmonary toxicity of ultrafine particles? II. Lung lavage studies.". *J Aerosol Science* 1990, 21: 384–387. 10.1016/0021-8502(90)90065-6
2. Kuhlbusch TAJ, Neumann S, Fissan H: Number size distribution, mass concentration and particle composition of PM1, PM2.5, PM10 in bag filling areas of carbon black production. *J Occup and Environ Hyg* 2004, 1: 660–671. 10.1080/15459620490502242
3. Kuhlbusch TAJ, Fissan H: Particle characteristics in the reactor and pelletizing area of carbon black production. *J Occup and Environ Hyg* 2006, 3: 558–567. 10.1080/15459620600912280
4. Maynard AD, Baron PA, Foley M, Shvedova AA, Kisin ER, Castranova V: Exposure to carbon nanotube material: Aerosol release during the handling of unrefined single walled carbon nanotube material. *J Toxic And Environ Health* 2004, 67: 87–107. Part A 10.1080/15287390490253688
5. ISO/TR 27628: Workplace atmospheres - Ultrafine, nanoparticle and nano-structured aerosols - Inhalation exposure characterization and assessment. 2007.
6. Szymczak W, Menzel N, Keck L: Emission of ultrafine copper particles by universal motors controlled by phase angle modulation. *J Aerosol Sci*2007,38(5):520–531. 10.1016/j.jaerosci.2007.03.002
7. Koponen IK, Jensen KA, Schneider T: Sanding dust from nanoparticle-containing paints: physical characterisation. *J Phys: Conf Ser* 2009, 151:012048.
8. Koponen IK, Jensen KA, Schneider T: Comparison of dust released from sanding conventional and nanoparticle-doped wall and wood coatings. *J Expo Sci Env Epid* 2010, 1–10.
9. Schneider T, Brouwer DH, Koponen IK, Jensen KA, Fransmann W, van Duuren-Stuurman B, van Tongeren M, Tielemans E: Conceptual model for assessment of inhalation exposure to manufactured nanoparticles. *J Exp Sci and Environ Epidem* 2011, 1–14.
10. Hoppel WA: Determination of the aerosol size distribution from the mobility distribution of charged fraction of aerosols. *Journal of Aerosol Science* 1978, 9: 41–54. 10.1016/0021-8502(78)90062-9

11. Wang SC, Flagan R: Scanning electrical mobility spectrometer. *Aerosol Science and Technology* 1990, 13: 230–240. 10.1080/02786829008959441
12. Tammet H, Mirme AA, Tamm E: Electrical aerosol spectrometer of Tartu University. *J Aerosol Sci* 1998,29(S1):S427-S428.
13. Tammet H, Mirme AA, Tamm E: Electrical aerosol spectrometer of Tartu University. *Atmos Res* 2002, 62: 315–24. 10.1016/S0169-8095(02)00017-0
14. Keskinen J, Pietarinen K, Lehtimäki M: Electrical low pressure impactor.*Journal of Aerosol Science* 1992, 23: 353–360.
15. Baron PA, Willeke K: *Aerosol Measurement - Principles, Techniques, and Applications.* New York: Wiley-Interscience; 2001.
16. Wilson JC, Liu BY: Aerodynamic particle size measurement by laser-doppler velocimetry. *J Aerosol Sci* 1980,11(2):139–150. 10.1016/0021-8502(80)90030-0
17. Berner A: Praktische Erfahrungen mit einem 20-Stufen-Impaktor. *Staub - Reinhaltung der Luft* 1972, 32: 315–320.
18. Berner A, Lürzer C, Pohl F, Preining O, Wagner P: The size distribution of the urban aerosol in Vienna. *The Science of the Total Environment* 1979, 13: 245–261. 10.1016/0048-9697(79)90105-0
19. Gorbunov B, Priest ND, Muir RB, Jackson PR, Gnewuch H: A Novel Size-Selective Airborne Particle Size Fractionating Instrument for Health Risk Evaluation. *Ann Occup Hyg* 2009, 53: 225–237. 10.1093/annhyg/mep002
20. Azong-Wara N, Asbach C, Stahlmecke B, Fissan H, Kaminski H, Plitzko S, Kuhlbusch TAJ: Optimisation of a thermophoretic personal sampler for nanoparticle exposure studies. *J Nanoparticle Research* 2009, 11: 1611. 10.1007/s11051-009-9704-0
21. McMurry PH: The history ofcondensation nucleus counters. *Aerosol Science and Technology* 2000, 33: 297–322. 10.1080/02786820050121512
22. Farnsworth J, Caldow R: A Method for Empirical Measurement of CPC Upper-end Counting Efficiency. In *International Aerosol Conference.* Helsinki; 2010.
23. Ku BK, Maynard AD: Comparing aerosol surface area measurement of monodisperse ultrafine silver agglomerates using mobility analysis, transmission electron microscopy and diffusion charging. *J Aerosol*

*Sci*2005,36(9):1108–1124. 10.1016/j.jaerosci.2004.12.003

24. Fissan H, Neumann S, Trampe A, Pui DYH, Shin WG: Rationale and principle of an instrument measuring lungdeposited nanoparticle surface area. *J Nanopart Res* 2007, 9: 53–59.
25. Shin WG, Pui DYH, Fissan H, Neumann S, Trampe A: Calibration and numerical simulation of nanoparticle surface area monitor (TSI model 3550 NSAM). *J Nanopart Res* 2007,9(1):61–69.
26. Dixkens H, Fissan H: Development of an Electrostatic Precipitator of Off-Line Particle Analysis. *Aerosol Science and Technology* 1999, 30: 438–453. 10.1080/027868299304480
27. ISO 15900: Determination of particle size distribution - Differential electrical mobility analysis for aerosol particles. 2009.
28. Asbach C, Fissan H, Stahlmecke B, Kuhlbusch TAJ, Pui DYH: Conceptual Limitations and Extensions of Lung Deposited Nanoparticle Surface Area Monitor (NSAM). *J Nanopart Res* 2009, 11: 101–109. 10.1007/s11051-008-9479-8
29. Stahlmecke B, Wagener S, Asbach C, Kaminski H, Fissan H, Kuhlbusch TAJ: Investigation of airborne nanopowder agglomerate stability in an orifice under various differential pressure conditions. *J Nanopart Res*2009,11(7):1625–1635. 10.1007/s11051-009-9731-x
30. Kuhlbusch TAJ, Fissan H, Asbach C: Measurement and Detection of Nanoparticels Within the Environment. In *Nanotechnology: Volume 2: Environmental Aspects*. Edited by: Krug H. Wiley-VCH, Weinheim; 2008:229–266.
31. McMurry PH: A review of atmospheric aerosol measurements. *Atmospheric Environment* 2000, 34: 1959–1999. 10.1016/S1352-2310(99)00455-0
32. Hinds WC: *Aerosol technology - properties, behaviour, and measurement of airborne particles.* New York: John Wiley and Sons; 1999.
33. Methner M, Hodson L, Dames A, Geraci C: Nanoparticle Emission Assessment Technique (NEAT) Part B. *J Occup and Environ Hyg* 2010, 7: 163–176.
34. Consent Report: Tiered approach to an exposure measurement and assessment of nanoscale aerosols released from engineered nanomaterials in workplace operations, BAuA, BGRCI, IFA and VCI, Germany. 2011, in press.

35. Maynard AD, Aitken RJ: Assessing exposure to airborne nanomaterials: Current abilities and future requirements. *Nanotoxicology* 2007,1(1):26–41. 10.1080/17435390701314720

36. Brouwer D, van Duuren-Stuurman B, Berges M, Jankowska E, Bard D, Mark D: From workplace air measurement results towards estimates of exposure? Development of a strategy to assess exposure to manufactured nano-objects. *J Nanopart Res* 2009, 11: 1867–1881. 10.1007/s11051-009-9772-1

37. Wang YF, Tsai PJ, Chen CW, Chen DR, Hsu DJ: Using a modified electrical aerosol detector to predict nanoparticle exposure to different regions of the respiratory tract for workers in Carbon Black manufacturing industry. *Environ Sci Technol* 2010, 44: 6767–6774. 10.1021/es1010175

38. Johnson DR, Methner MM, Kennedy AJ, Steevens JA: Potential for Occupational Exposure to Engineered Carbon-Based Nanomaterials in Environmental Laboratory Studies. *Environ Health Perspect* 2010,118(1):49–54.

39. Demou E, Stark WJ, Hellweg S: Particle emission and exposure during nanoparticle synthesis in research laboratories. *Ann Occup Hyg* 2009, 53:829–838. 10.1093/annhyg/mep061

40. Manodori L, Benedetti A: Nanoparticles monitoring in workplaces devoted to nanotechnologies. *J Phys Conf Ser* 2009, 170: 012001.

41. Park J, Kwak BK, Bae E, Lee J, Kim Y, Choi K, Yi J: Characterization of exposure to silver nanoparticles in a manufacturing facility. *J Nanopart Res* 2009, 11: 1705–1712. 10.1007/s11051-009-9725-8

42. Bello D, Wardie BL, Yamamoto N, deVilloria RG, Garcia EJ, Hart AJ, Ahn K, Ellenbecker MJ, Hallock M: Exposure to nanoscale particles and fibres during machining of hybrid advanced composites containing carbon Nanotubes. *J Nanopart Res* 2009,5(1):231–249.

43. Möhlmann C, Welter J, Klenke M, Sander J: Workplace exposure at nanomaterial production processes. *J Phys Conf Ser* 2009, 170: 012004.

44. Helsper C, Horn HG, Schneider F, Wehner B, Wiedensohler A: Intercomparison of five mobility particle size spectrometers for measuring atmospheric submicrometer aerosol particles. *Gefahrst Reinhalt Luft* 2008, 68: 475–481.

45. Demou E, Peter P, Hellweg S: Exposure to manufactured nanostructured particles in an industrial pilot plant. *Ann Occup Hyg* 2008,52(8):695–706. 10.1093/annhyg/men058

46. Yeganeh B, Kull CM, Hull MS, Marr LC: Characterization of Airborne Particles during production of carbonaceous nanomaterials. *Environ Sci Technol* 2008, 42: 4600–4606. 10.1021/es703043c

47. Peters T, Elzey S, Johnson R, Park H, Grassian V, Maher T, O'Shaughnessy P: Airborne Monitoring to Distinguish engineered Nanomaterials from Incidental particles for environmental health and safety. *J Occup and Environ Hyg* 2009, 6: 73–81. 10.1080/15459620903249695

48. Fujitani Y, Kobayashi T, Arashidani K, Kunugita N, Suemura K: Measurement of physical properties of aerosols in a fullerene factory for inhalation exposure assessment. *J Occup and Environ Hyg* 2008, 5: 380–389. 10.1080/15459620802050053

49. Tsai SJ, Hofmann M, Hallco M, Ada E, Kong J, Ellenbecker M: Characterization and evaluation of nanoparticle release during the synthesis of single-walled and multiwalled carbon nanotubes by chemical vapor deposition.*Env Sci Technol* 2009, 43: 6017–6023. 10.1021/es900486y

50. Evans DE, Ku BK, Birch ME, Dunn KH: Aerosol monitoring during carbon nanofiber production: Mobile direct-reading sampling. *Ann Occup Hyg*2010,54(5):514–531. 10.1093/annhyg/meq015

51. Wang J, Asbach C, Fissan H, Huelser T, Kuhlbusch TAJ, Thompson D, Pui DYH: How can nanobiotechnology oversight advance science and industry: Examples from environmental, health and safety studies of nanoparticles (nan-EHS). *J Nanopart Res* 2011, 13: 1373–1387. 10.1007/s11051-011-0236-z

52. Mazzuckelli LF, Methner MM, Birch ME, Evans DE, Ku BK, Crouch K, Hoover MD: Identification and characterization of potential sources of worker exposure to carbon nanofibres during polymer composite laboratory operations. *J Occup and Environ Hyg* 2007, 4: D125-D130. 10.1080/15459620701683871

53. Tsai CJ, Wu CH, Leu ML, Chen SC, Huang CY, Tsai PJ, Ko FH: Dustiness test of nanopowders using a standard rotating drum with a modified sampling train. *J Nanopart Res* 2009,11(1):121–131. 10.1007/s11051-008-9453-5

54. Bello D, Hart AJ, Ahn K, Hallock M, Yamamoto N, Garcia EJ, Ellenbecker MJ, Wardle BL: Particle exposure levels during CVD growth and subsequent handling of vertically-aligned carbon nanotube films. *Carbon* 2008, 46:974–981. 10.1016/j.carbon.2008.03.003

55. Tsai SJ, Ashter A, Ada E, Mead JL, Barry CF, Ellenbecker MJ: Control of Airborne Nanoparticles Release During Compounding of Polymer Nanocomposites. *Nano* 2008,3(4):301–309. 10.1142/S179329200800112X

56. Kuhlbusch TAJ, Kaminski H, Jarzyna D, Fissan H, Asbach C: Measurements of nanoscale TiO_2 and Al_2O_3 in workplace environments - Methodology and results. *JOEH* 2011, in press.

57. Oberdörster G: Toxicology of ultrafine particles. In vivo studies. *Phil Trans R Soc Lond A* 2000, 358: 2719–2740. 10.1098/rsta.2000.0680

58. Brown DM, Wilson MR, MacNee W, Stone V, Donaldson K: Size-Dependent Proinflammatory Effects of Ultrafine Polystyrene Particles: A Role for Surface Area and Oxidative Stress in the Enhanced Activity of Ultrafines.*Toxicology and Applied Pharmacology* 2001, 175: 191–199. 10.1006/taap.2001.9240

59. Duffin R, Tran CL, Clouter A, Brown DM, MacNee W: The Importance of Surface Area and Specific Reactivity in the Acute Pulmonary Inflammatory Response to Particles. *Annals of Occupational Hygiene 2002*46(Suppl 1):242–245.

60. Asbach C, Kaminski H, Fissan H, Monz C, Dahmann D, Mülhopt S, Paur HR, Kiesling HJ, Herrmann F, Voetz M, Kuhlbusch TAJ: Comparison of four mobility particle sizers with different time resolution for stationary exposure measurements. *J Nanopart Res* 2009,11(7):1593–1609. 10.1007/s11051-009-9679-x

61. Tsai SJ, Ashter A, Ada E, Mead JL, Barry CF, Ellenbecker MJ: Airborne nanoparticle release associated with the compounding of nanocomposites using nanoalumina as fillers. *Aerosol Air Qual Res* 2008, 8:160–177.

62. Lee JH, Kwon M, Ji JH, Kang CS, Ahn KH, Han JH, Yu IJ: Exposure assessment of workplaces manufacturing nanosized TiO_2 and silver. *Inhal Toxicol*2011,23(4):226–236. 10.3109/08958378.2011.562567

63. Guiot A, Golanski L, Tardif F: Measurement of nanoparticle removal by abrasion. *J Phys: Conf Ser* 2009, 170: 012014.

64. Vorbau M, Hillemann L, Stintz M: Method for the characterization of the abrasion induced nanoparticles release into air from surface coatings. *J Aerosol Sci* 2009,40(3):209–217. 10.1016/j.jaerosci.2008.10.006

65. Schneider T, Jensen KA: Combined Single-Drop and Rotating Drum Dustiness Test of Fine to Nanosize Powders Using a Small Drum. *Ann Occup Hyg* 2008,52(1):23–34.

66. Jensen KA, Koponen IK, Clausen P, Schneider T: Dustiness behaviour of loose and compacted Bentonite and organoclay powders: What is the difference in exposure risk? *J Nanopart Res* 2009,11(1):133–146. 10.1007/s11051-008-9420-1

67. Göhler D, Stintz M, Vorbau M, Hillemann L: Characterization of nanoparticle release from surface coatings by the simulation of a sanding process. *Ann Occup Hyg* 2010,54(6):615–624. 10.1093/annhyg/meq053

68. ISO/DIS 12025: Nanomaterials - Quantification of nano-object release from powders by generation of aerosols. 2010.

69. Bello D, Wardle BL, Zhang J, Yamamoto N, Santeufemio C, Hallock M, Virji MA: Characterization of exposures to nanoscale particles and fibers during solid core drilling of hybrid CNT advanced composites. *Int J Occup Environ Health* 2010, 16: 434–450.

70. Isbaseta N, Biscans B: Ultrafine Aerosol Emission from Free Fall of TiO2 and SiO2 Nanopowders. *Kona* 2007, 25: 190–204.

71. Maynard AD: Experimental Determination of Ultrafine TiO_2 Deagglomeration in a Surrogate Pulmonary Surfactant: Preliminary Results. *Ann Occup Hyg* 2002,46(S1):197–202.

72. Baron PA, Maynard AD, Foley M: Evaluation of Aerosol Release during the Handling of Unrefined Single Walled Carbon Nanotube Material. 2003.

73. Ogura I, Sakurai H, Gamo M: Dustiness testing of engineered nanomaterials. *J Phys: Conf Ser* 2009, 170: 012003.

74. Lee SB, Lee JH, Bae GN: Size response of an SMPS-APS system to commercial multi-walled carbon nanotubes. *J Nanopart Res* 2010,12(2):501–512. 10.1007/s11051-009-9745-4

75. Plitzko S, Gierke E, Dziurowitz N, Broßell D: Generation of CNT/CNF dusts by a shaker aerosol generator in combination with a thermal precipitator as the collection system for characterization of the fibre morphology.*Gefahrstoffe - Reinhalt Luft* 2010,70(1–2):31–35.

76. Boundy M, Leith D, Polton T: Method to Evaluate the Dustiness of Pharmaceutical Powders. *Ann Occup Hyg* 2006,50(5):453–458. 10.1093/annhyg/mel004

77. Hagendorfer H, Lorenz C, Kaegi R, Sinnet B, Gehrig R, Goetz NV, Scheringer M, Ludwig C, Ulrich A: Size-fractionated characterization and quantification of nanoparticle release rates from a consumer spray product containing engineered nanoparticles. *J Nanopart Res* 2010,12(7):2481–2494. 10.1007/s11051-009-9816-6

78. DIN 68861–2: Furniture surfaces - Behaviour at abrasion. 1981.

79. DIN EN 13523–16: Coil coated metals - Test methods - Part 16: Resistance to abrasion. 2005.

80. DIN EN ISO 7784: Paints and varnishes - Determination of resistance to abrasion. 2006.

81. Hsu L, Chein H: Evaluation of nanoparticle emission for TiO2 nanopowder coating materials. *J Nanopart Res* 2007,9(1):157–163.

82. Wohlleben W, Brill S, Meier MW, Mertler M, Cox G, Hirth S, von Vacano B, Strauss V, Treumann S, Wiench K, Ma-Hock L, Landsiedel R: On the Lifecycle of nanocomposites: Comparing Released Fragments and their In-Vivo Hazards from Three Release Mechanisms and Four Nanocomposites.*Small* 2011.

83. Shi T, Schins RPF, Knaapen AM, Kuhlbusch TAJ, Pitz M, Heinrich J, Borm PJA: Hydroxyl radical generation by electron paramagnetic resonance as a new method to monitor ambient particulate matter composition. *J Environ Monit* 2003, 5: 550–556. 10.1039/b303928p

84. ISO/TS 27687: Nanotechnologies - Terminology and definitions for nano-objects - Nanoparticle, nanofibre and nanoplate. 2008.

11

Prospects of using Nanotechnology for Food Preservation, Safety, and Security

Vivek K. Bajpai [a], Madhu Kamle [b], Shruti Shukla [a], Dipendra Kumar Mahato [c], Pranjal Chandra [d], Seung Kyu Hwang [e], Pradeep Kumar [b], Yun Suk Huh [e], Young-Kyu Han [a]

[a] Department of Energy and Materials Engineering, Dongguk University-Seoul, 30 Pildong-ro 1-gil, Seoul, 04620, South Korea
[b] Department of Forestry, North Eastern Regional Institute of Science and Technology, Nirjuli, 791109, Arunachal Pradesh, India
[c] Department of Agriculture and Food Engineering, Indian Institute of Technology Kharagpur, West Bengal, 721302, India
[d] Department of Biosciences and Bioengineering, Indian Institute of Technology Guwahati, Guwahati, 781039, Assam, India
[e] Department of Biological Engineering, Biohybrid Systems Research Center (BSRC), Inha University, 100 Inha-ro, Nam-gu, Incheon, 22212, South Korea

Citation: Vivek K.Bajpai, Madhu Kamle, Shruti Shukla, Dipendra Kumar Mahato, Pranjal Chandra, Seung Kyu Hwang, Pradeep Kumar, Yun Suk Huh, Young-Kyu Han, 2018, Prospects of using nanotechnology for food preservation, safety, and security, Open Access Journal of Food and Drug Analysis, Volume 26, Issue 4, October 2018, Pages 1201-1214, https://doi.org/10.1016/j.jfda.2018.06.011.

ABSTRACT

The rapid development of nanotechnology has transformed many domains of food science, especially those that involve the processing, packaging, storage, transportation, functionality, and other safety aspects of food. A wide range of nanostructured materials (NSMs), from inorganic metal, metal oxides, and their nanocomposites to nano-organic materials with bioactive agents, has been applied to the food industry. Despite the huge benefits nanotechnology has to offer, there are emerging concerns regarding the use of nanotechnology, as the accumulation of NSMs in human bodies and in the environment can cause several health and safety hazards. Therefore, safety and health concerns as well as regulatory policies must be considered while manufacturing, processing, intelligently and actively packaging, and consuming nano-processed food products. This review aims to provide a basic understanding regarding the applications of nanotechnology in the food packaging and processing industries and to identify the future prospects and potential risks associated with the use of NSMs.

Keywords: Nanostructured materials (NSMs), Food safety, Food preservation, Functional food, Packaging, Human health

INTRODUCTION

Nanotechnology integrates several disciplines, including physics, chemistry, biotechnology, and engineering, and refers to the use of nanomaterials whose nanoscale structures range from 1 to 100 nm [1]. In these nano-sizes, materials adopt unique properties that were not present when the materials were in their original form. Understanding these unique properties in order to develop new and improved products through green processes is the goal of nano-scientists across the globe [2].

Nanomaterials or nanostructure materials exhibit different dimensions for their structural elements, clusters, molecules, or crystallites, such as zero dimensions (nanoparticles, nanoclusters, and quantum dots), one dimension (nanorods and nanotubes), two dimensions (nano-thin films), and three dimensions (nanomaterials) in the 1–100 nm range [3], [4]. Combining nanostructure materials with other polymer, biomolecule, and other nanostructure material or existing in the aggregate form can result relatively in a larger particle size material (>100 nm) leading to the formation of nanocomposite [5]. These nanomaterials having high surface volume ratio exhibit inimitable physio-chemical characteristics, such as solubility, toxicity, strength, magnetism, diffusivity, optics, color, and thermodynamics [6], [7].

The applications of these nanomaterials are growing in various sectors, including agriculture, medicine, clothing, cosmetics, food, and public health due to their unique ability to increase solubility and bioavailability and to protect bioactive components while they are being processed and stored [4], [8], [9]. Due to the excellent physiochemical nature and the antimicrobial potential of nanomaterials, they are widely used against various pathogenic microbes and in healthcare, crop protection, water treatment, food safety, and food preservation [8], [10]. In addition, nanostructured materials (NSMs) are being applied in the food industry as a nanosensor, new packaging material, and encapsulated food component. In general, this review concerns the use of nanotechnology, specifically nanoparticles and constructed nanostructures, in the food sector.

Nanotechnology has a wide range of food-related applications. In these applications, a specific type of nanomaterial is incorporated into a specific food product in order for that food product to develop certain desired properties [2]. The field of nanotechnology has also been an integral part of research and development for the large-scale manufacturing of agricultural products and processed foods and drinks, as well as for food packaging across the world [3]. Several reports confirm that these nanomaterials can successfully improve food safety by enhancing the efficacy of food packaging, shelf-life, and nutritional value as additives without changing the taste and physical characteristics of food products [13], [14]. Although they have much potential to frame innovative products and production processes in the food industry, nanotechniques are facing a major challenge in using cost-effective processing operations to create edible and non-toxic nano-delivery systems and to develop effective formulations that are safe for human consumption. Therefore, due to the increased use of these NSMs, there have been growing concerns regarding the development of biocompatible, safe, and non-toxic nanostructures from food-grade ingredients using simple, green, and cost-effective strategies, including the layer-by-layer technique [15].

United states-based safety evaluation agencies such as FDA and EPA, including the Directorate of European Health and Consumer Protection, and many other regulatory authorities, have released several guidelines on potential human health risks associated with the use of nanostructure materials in food [16]. These potential risks are determined by nanoscientists using nanotoxicology, which is a branch of toxicology and an interdisciplinary field that concerns varied toxicity aspects of nanomaterials [15]. Nanotoxicity mediated via generation of reactive oxygen species (ROS) is considered a crucial mechanism leading to successive human oxidative stress [8]. Also, there is a huge interest in using nanomaterials in various fields, including their safety concerns. Consequently, there is a compelling demand to address these issues so as to expand our knowledge on the use of nanomaterials/ nanostructures in terms of their biocompatibility, safety, and toxicity in the food sector.

Although nanotechnology has applied green-synthesized NSMs in several ways in the food sector, on a few occasions, the use of NSMs has been controversial, as NSMs are scientifically uncertain and can impart long-term detrimental effects on human health and the environment [17]. In this regard, the complexity and limitations in the field of nanotechnology in terms of toxicity and accumulation (in appropriate doses) can be overcome only by elucidating the physiochemical and biological properties of NSMs through extensive research [18]. In light of the effects of nanomaterials on the food industry, this review has been formulated as a "snapshot" to address the current status of the techniques and implementations of nanotechnology in food preservation, safety, and security criteria and to identify the underlining mechanistic actions regarding safety or packaging issues. In addition, the review focuses on the current and future prospects of nanotechnology and discusses the strengths and weaknesses (i.e. the risks related to toxicological effects) of existing regulatory authorities. Moreover, the review discusses the current efforts to address these weaknesses and other issues related to the development, understanding, and promotion of nanotechnology. Fig. 1illustrates an overview of the important usages of nanoparticles/ nanomaterials and nanostructures in the food sector.

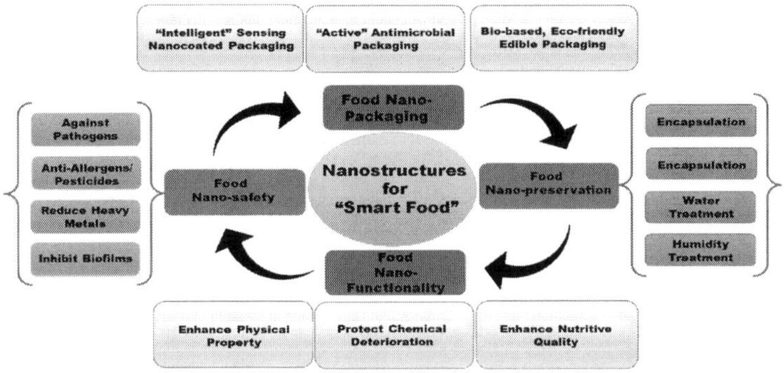

Figure 1: Systematic representation of application of nanoparticles in various areas of food industry.

NATURAL AND SYNTHETIC NANOSTRUCTURES IN THE FOOD SYSTEM

The food system naturally contains a variety of nano-sized elements that include self-assembled higher order structures, such as carbohydrates and fats [19]. These ingredients are different from synthetically manufactured nanomaterials/nanostructures thus can be used to make nanoemulsion, nano-encapsulate, and food-grade polymer [19]. During food processing, manufacturing, or thermal treatments, nanostructures are used to produce, which are not associated with modern nanotechnology (e.g. coagulation, emulsifying, or homogenizing). In brief, food proteins, including polysaccharides and lipids, are globular particles whose size can vary between 10 and several hundred nanometers. Coagulation and emulsification are based on the formation of reticular, zero, one, two and three-dimensional nanostructures. When corn starch is boiled to make custard-like food, small, three-dimensional crystalline structures that are only tens of nanometers in thickness are melted [20]. Fresh milk and milk-based products naturally contain nanostructures, such as milk proteins and casein. When milk is homogenized, fat globules that are about 100 nm in size are produced [20].

Nanotechniques play an essential role in the functional and nutraceutical food sectors. Encapsulation of effective molecules or components in NSMs can enhance the bio-availability and solubility of coloring agents and nutritional ingredients such as minerals and vitamins, thus facilitating controlled release and protecting biologically active substances and micro-nutrients during their processing. Currently, the most usable NSMs are nanocapsules (e.g. micellesand liposomes) and nanoemulsions, including a few carbon-based, green-synthesized, ecofriendly nanomaterials.

NANOTECHNOLOGY IN FOOD PACKAGING AND SECURITY

The packaging of food is one of the most critical steps in terms of food safety. Natural substances, atmospheric gases and water vapors are absolutely impermeable to no packaging materials [21], [22]. However, completing blocking the migration and permeability of gases is not desirable in the case of packaging fresh fruits and vegetables that undergo cellular respiration [21]. In contrast, the packaging of carbonated beverages should eliminate the flow of oxygen and carbon dioxide (CO_2) to prevent oxidation and de-carbonation [21]. The flow of CO_2, oxygen, and water vapors varies according to the food matrices and the packaging materials that are used. Hence, these complexities in food packaging can be addressed and overcome by employing various nanocomposite materials, including polymers [24]. Nanoparticle having a diameter of less than 100 nm is thousand-times thinner than about 100,000 nm thick book page or hundred-times thinner than about 10,000 nm thick human hair. With an appropriate regulatory processing approval, different structures of nanoparticles can be of significant use in various fields of food and pharmaceutical industries, especially for the research programs associated with food science-based research (Fig. 2).

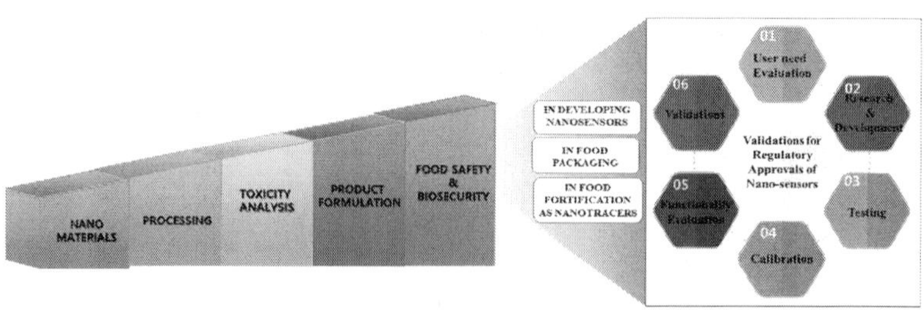

Figure 2: Steps for processing and utilizing of nanomaterials in food sectors.

Recently, the use of nano-biocomposites in food packaging has enhanced the ability of food packaging to act as a barrier against gases [25]. The applications of various nanoparticles in food packaging are briefly shown in Table 1. Recent trends in food packaging are encouraging the use of biodegradable polymers reinforced with nanofillers, which are eco-friendly [26]. However, there is a major concern regarding the ingestion of these nano-compounds during consumption of the food. Hence, it is essential to investigate how these nanoparticles migrate within the human body as well as these nanoparticles' toxic and immunogenic effects [27]. Another concern is related to the biodegradability of these nanofilled, biodegradable polymers [28]. These concerns are seriously considered by researchers across the world who are searching for human and environmentally friendly nanomaterials [29].

Table 1: Nanoparticles for application in food packaging

Types of nanoparticles	Matrix	Application	Reference
Silver	Asparagus, Orange juice, Poultry meat, Fresh-cut melon, Beef meat exudates	Retards the growth of aerobic psychrotrophics, yeasts and molds; antimicrobial effect against *Escherichia coli*and *Staphylococcus aureus*	[165], [166], [167], [168], [169]
Zinc oxide	Orange juice, Liquid egg albumen	Effectively reduces *Lactobacillus plantarum*, *Salmonella*, yeast and mold counts without changes in quality parameters	[169], [170], [171]
Titanium oxide	Chinese jujube, Strawberry	Reduces browning, slow-down ripening, senescence and decay	[172]
Silver oxide	Apple slice	Retards microbial spoilage	[173]

Nanocoatings as Intelligent Packaging for Surfaces

In food packaging, NSMs, with biopolymers, can either enhance the property characteristics of neat polymer or improve the functional properties of active and intelligent packaging. Packaging can be categorized as "Improved", "Active," and "Intelligent" packaging, which represent which kind of packaging material could be used for what applications [30]. However, official restrictions have been imposed by the European Union (EU) using "active" and "intelligent" packaging materials in foods except using titanium nitridein plastic bottles [31].

For in-packaging detection of oxygen, Mill [32] developed nano-sized titanium dioxide (TiO_2) or tin dioxide particle-based favorable photo-indicator intelligent ink using a redox-activate methylene blue dye, and this detector was gradually able to change the color in response to minor changes in oxygen quantity. In recent years, the packaging industry has focused on the various types of nanostructures and specified nanomaterials, including nano-clay particulates due to their easy accessibility, low-cost, easy process ability, and great performance. In addition, graphene nano-sheets and carbon nano-tubes are also encouragingly developed as a carbon-based nanomaterial [33]. Polymeric substrate-based flexible or static films and bottles (various materials, such as glass and plastic etc.) in combination with vacuum-coated thin inorganic layers have great demand in material packaging due to their ability to block out oxygen and aromas. While aluminum foil and aluminum-metallization used to be the material and method of choice, respectively, new materials are being used more frequently.

Thick and uniform layers (10–100 nm coating thickness) are broadly referred to as "nanocoatings" [34]. In recent years, few researchers have developed an intelligent type of nanocoating film that can indicate whether any contamination has occurred during storage. Gas content and non-invasive detection methods have also shown great ability to continually and easily monitor the gas content, excess moisture, and oxygen content of a package-headspace thus providing effective means to evaluate the quality and safety of food even after the production process [35], [36]. Presence of oxygen inside the packaging can cause shelf-life threatening of food due to the ability of oxygen to create healthy environment for microbial growth [37]. Yu et al. [38] developed polyvinyl alcohol/chitosan polymer-based biodegradable and cost-effective films using a nanocomposite of silica *in-situ* which caused significant reduction in the permeability of oxygen and moisture by 25.6% and 10.2%, respectively, suggesting that the films extended the preservation time of cherries by three-fold compared to normal packaging. Swaroop and Shukla [39] developed a food packaging material using a combination of nanostructures of magnesium oxide (MgO) and polylactic acid biopolymer, and found that the material effectively protected against bacterial biofilms. Furthermore, Foltynowicz et al. [40] synthesized zero-valent iron particles to act as oxygen scavengers in food packaging.

The Mechanisms of Nanoparticles/Nanocapsules as Antimicrobial Agents for Active Packaging

The outstanding use of nanoparticles with antimicrobial properties has, to some extent, protected our food products from foodborne illness outbreaks due to the consumption of spoiled packaged food. When compared with conventional food-packagings, an active food-packaging not only has the ability of being a passive barrier, but also helps to release antioxidant and antimicrobial compounds through the direct interaction with the food as well as helps to remove some negative factors, such as oxygen or water vapor. These interactions usually improve food stability [41].

Active packaging contains specific molecules which have the ability to absorb or release the components into or from the surrounding environment of the packaged food. Today, active polymer nanomaterials in food packaging have mainly been developed for antimicrobial packaging applications [42]. Overall, in order to improve the functionality and effectiveness of food packages, various bioactive substances can be incorporated into the packaging material by capsulation, capping the material with nanomaterials, or employing other nanotechniques [43]. To effectively apply antimicrobial active packaging to food products, food providers should select effective antimicrobial substances-based unique methodology to ensure the quality of the packaged products, such as their visual appearance and sensory levels [44].

The encapsulation of nutraceuticals and functional antimicrobial ingredients is essential for the preservation and bioavailability of bioactive ingredients, as it is extremely important in food processing, food storage, and passage through the gastrointestinal tract [5]. The macromolecule-based nanoparticles in food not only improve the bioavailability of bioactive polyphenolics, such as resveratrol, epigallocatechin-3-gallate, and curcumin, but also enhance the solubility of these polyphenols and thus prevent their degradation in the gastrointestinal environment [9]. There are various encapsulation techniques that have been used to produce nano or micro-particulate systems, such as nano-emulsion [45], coacervation [46], [47], the extrusion method [48], fluidized bed coating [49], spray cooling [50], and spray drying [51].

Nano-encapsulation allows for the direct contact of nanomaterials during the consumption of food. In a commonly known example, silicon dioxide (SiO_2) is widely applied as a fragrance carrier in various food products [52]. Several lipid-based, nano-encapsulation systems have been developed to increase the activity of antioxidant components by improving their bio-availability and solubility [53] as well as their targeted, site-specific delivery with efficient absorption [54]. Studies have found that nano-sized edible coating is an effective alternative for preservation and shelf-life extension of food, as well as prevention of food from microbial spoilage [55]. Efficacy of coatings made with gelatin incorporating nano-crystals of cellulose [56], chitosan or nano-silica [57], chitosan with nano-silica [58], and alginate or lysozyme nano-laminate has been found very effective for the preservation of the quality of fresh foods during prolonged storage [59]. Additionally, the nano-packaging-based method that incorporates blended polyethylene with nano-powders such as Ag, kaolin, anatase TiO_2, and rutile TiO_2 has been considered to be a novel and facile technique to preserve fruits, such as strawberries (*Fragaria ananassa* Duch.) [60]. Despite the tremendous research in nano-encapsulation using various materials [61], the precise delivery of foods and their safety has not been studied in detail. Therefore, future research should examine the long-term toxicity of nano-encapsulated foods [62].

In addition to this, Johnston [63] employed nanostructured calcium silicate(NCS) for the absorption of Ag^+ ions from a solution, which formed an NCS–Ag complex that was successfully used as an antimicrobial agent for food packaging purposes. Similarly, TiO_2 is commonly utilized in surface-coating as a photocatalytic disinfecting material [64] and to inactivate several pathogenic bacteria in food [65]. The antibacterial potential of chitosan-capsulated nanoparticles could be mediated through the interaction between the positively-charged chitosan and the negatively-charged cell membranes. These interactions lead to enhanced membrane permeability and cause the rupture and eventual leakage of the intracellular material [66].

Sarwar et al. [67] developed polyvinyl alcohol (PVA), nanocellulose and Ag nanocomposite films and applied them to antimicrobial food packaging. These films profound antimicrobial effects against both Methicillin Resistant *Staphylococcus aureus* (MRSA) and DH5-alpha strain of *Escherichia coli*. Valerini et al. [68] also developed nanostructured, and aluminum-doped zinc oxide-based antimicrobial coatings functionalized with polylactic acid using sputtering power. The coatings exhibited strong antibacterial potential against *E. coli* and were proposed as promising materials for environment-friendly active packaging. Moreover, Lu et al. [69] prepared an antimicrobial nano-emulsion that was 100 nm by encapsulating citral essential oil using ultrasonic power, and as a result, the designed nano-emulsion was significantly effective in delivering an antimicrobial agent (citral essential oil) in the food system.

Metal and metal oxide nanoparticles (NPs)-based active packagings could be effective tools of nano-composites in terms of their antimicrobial potential. Among all metal nanoparticles, silver (Ag^+) nanoparticles are widely used in commercial sectors. Silver nanoparticles are toxic for various food pathogens[70] because silver nanoparticles increase cell membrane permeability via cell surface adherence and by degrading lipopolysaccharide [71]; in this way, the silver nanoparticles can penetrate the bacterial cell, damage its DNA [72], and release antimicrobial Ag^+ ions [73], which bind to sulfur, oxygen, or nitrogencontaining electron donor groups in molecules and thus inhibiting synthesis of adenosine triphosphate (ATP) and DNA replication, eventually leading the cell to die [74]. Millimolar concentrations of Ag^+ ions can easily cause cytoplasm shrinkage; separation of the cell wall membrane; the devastation of the peptidoglycan in the cell wall; denaturation of ribosomes; and DNA compression, which inhibits DNA synthesis, ruptures the cell membrane, and causes cell death [74]. Basically, the three major mechanisms of the bacterial toxicity of metal-containing NPs that are widely accepted include the uptake of metal ions, which causes the intracellular ATP depletion; ROS production, which causes oxidative cellular damage, and disruption of bacterial membrane [4].

Similarly, reports have confirmed incorporation, coating, immobilization and surface modification of antimicrobial components onto packaging material [75]. Also, the higher antimicrobial effect can be achieved via direct incorporation of antimicrobial components into packaging films. Similarly, films coated with various effective antimicrobial components can result in higher antimicrobial potential. While incorporating these antimicrobial agents (volatile and non-volatile) into film packaging, these molecules can be dispersed onto the food surfaces via adopting migration, evaporation or diffusion abilities [76]. Recently, Arfat et al. [77] designed a bio-nanocomposite film based on fish skin gelatin and silver-copper bimetallic NPs. The film showed significant antibacterial potential against *Listeria monocytogenes* and *Salmonella* enterica. Several silver-based, antimicrobial master batches, such as Bactiblock®, Aglon®, Surfacine®, d2p®, IonPure®,

Irgaguard®, and Biomaster®have been used to retard the growth of *Campylobacter*, *Salmonella*, and *E. coli*.

NANOTECHNOLOGY IN FOOD FUNCTIONALITY

Nanoparticles for Protection from Chemical Deterioration

The main reason behind deterioration in the quality of food lies in an array of chemical reactions that occur between various components present in food and the external environment. Researchers have identified several nanomaterials that can control these unwanted reactions in various food matrices. However, some of these metal and metal oxide nanomaterials cause nano-toxicity, primarily because they can form ROS and cause oxidative stress, which leads to an imbalanced redox state in the cell [78]. Therefore, relatively less reactive nanomaterials are employed as an antioxidant carrier [18], [79], [80]. Polymeric nanoparticles are suitable for the encapsulation of bioactive compounds, such as vitamins and flavonoids, and for the release of these compounds in an acidic environment, such as in the stomach [81]. The application of antioxidants and edible coatings control the browning, the conversion of phenolics into dark-colored pigments in the presence of oxygen of fresh-cut fruits [82]. Although few nanomaterials have been used directly as anti-browning agents, nano-ZnO has been used as coated active packaging to improve the shelf-life of fresh-cut Fuji apples [83].

Nanotechnology can enhance food functionality either by eliminating chemical toxicants or by enhancing nano-sized nutritional supplements. For instance, some of the nutraceuticals incorporated in the carriers consist of lycopene, β-carotenes, and phytosterols in order to avoid cholesterol accumulation in body [84]. It is well-known that food itself has few nanostructures that are responsible for self-mediated nano-effects. For example, green tea has many health benefits because of its nano-selenium content, which facilitates the effective uptake of selenium.

Nano-encapsulation involves a nanoscale process of packing material by using nanocapsules, and it ensures the functionality of the final product, which includes the controlled release of the core. Therefore, encapsulated forms of ingredients exhibit numerous advantages, including consecutive delivery of multiple active ingredients, shelf-life extension, increased stability, and pH-triggered controlled release [85]. Further, Liang et al. [86] encapsulated epigallocatechin gallate (EGCG in zein/chitosan) nanoparticles for controlled, effective, functional applications in the food system. It was observed that the release of EGCG from zein/chitosan (CS) NPs and the DPPH scavenging ability of the zein/CS NPs was relatively higher than that of the nanoparticles that did not have the zein/CS coating in fatty simulant of 95% ethanol. These findings confirmed that the antioxidant activity and controlled-release of EGCG from zein/CS NPs in fatty simulant of 95% ethanol protect fatty foods from chemical deteriorates via enhancing antioxidant efficiencies and thus may help fatty foods to protect from oxidation for longer period.

Nanoparticles for Enhancing the Physical Properties of Food and Packaging Materials

Nanomaterials developed have shown enormous ability to increase the physical properties of both food and packaging materials [87]. In the 1990s, polymer nanocomposites with layered silicates that had several important characteristics, such as high flame resistance [88] and protection from UV rays, were developed [89]. Moreover, a number of NPs have been developed to increase the physical attributes of food, such as color and appearance. $TiO_2$was approved as a coloring food additive by the USFDA with the agreement that the limit of TiO_2 should not exceed 1% (w/w) as an additive ingredient [90]. Besides TiO_2, mixture of color additives may contain SiO_2 and/or Al_2O_3, but the use of carbon black is strictly prohibited by the USFDA as a food coloring additive [91]. On the other hand, SiO_2 is used as an anti-caking agent to maintain the flow properties in powdered products and as a carrier of aroma in food and non-food products. Nano-sized SiO_2 is widely employed in food products and has been registered within the EU as a food additive, as per Directive E551 [52].

β-Carotene is a nutraceutical component that acts as a coloring agent and as provitamin A, but its incorporation into foods is limited because of its low chemical stability and hydrophobicity. To improve the physicochemical stability of β-carotene, Mehrad et al. [92] synthesized nanomaterial by encapsulating it into solid lipid nanoparticles (SLNPs) that contained palmitic acid and corn oil and stabilized the material using a whey protein isolate (WPI). In this nanostructure, the encapsulated β-carotene was protected by a solid shell made by palmitic acid crystals covering the surface of the oil droplets. Corn oil was able to decrease the exclusion of β-carotene from the solid lipid matrix to the surface of the SLNPs, while

WPI increased the stability of the colloidal system, thus improving the oxidative stability of the β-carotene.

NANOTECHNOLOGY IN FOOD SAFETY

Globally, food safety is a growing public health concern. The primary goal of food safety is to ensure that food will not cause any harm to the consumer during preparation and consumption [93]. Food must be protected from any type of physical, chemical, and biological contamination during processing, handling, and distribution [94]. Recent developments in nanotechnology have revolutionized the food industry with its various applications in food processing, safety, and security, as well as its strides in enhancing nutraceutical value, extending shelf-life, and reducing packaging waste [94]. Today, food safety is a major concern due to rapid changes in food recipes and food habits. Foodborne pathogens, toxins, and other contaminants can cause serious threats to human health. The conventional methods for the detection of pathogens and their toxins are labor intensive and time consuming. Advances in nanotechnology have expedited the addressing of food safety issues in microbial contaminants and have improved toxin detection, shelf-life, and packaging strategies [84]. In addition, nanomaterials, including metal nanoparticles, carbon nanotubes, quantum dots, and other active nanomaterials can be used to develop biosensors for the quantification of microbes and other tests for food safety applications [84], [94].

Nanoparticles for the Detection of Foodborne Pathogens

Nano-biosensors are bioanalytical devices that are developed using various NSMs and biological receptors in an integrated system design [96]. Various types of biosensors have been developed to detect foodborne pathogens and food spoiling materials [97], [98]. Surface enhanced Raman scattering (SERS) is employed as a nano-biosensing tool to rapidly and accurately detect microbial pathogens [99], [100]. Silver nano-colloids are commonly used in SERS [101] to detect bacteria, as silver nano-colloids enhance Raman signals. Besides silver nano-colloids, graphene oxide [102], magnetic beads [103], carbon nanotubes [104], plasmonic gold [44], and silver nanoparticles [24] are commonly used to detect food-associated bacterial pathogens. In addition, synthetic DNAmolecular beacons tagged with color-coded probes are used as nano-barcodes to identify food pathogens [72].

The direct detection of *E. coli* in food samples has also become possible by the measurement and detection of light that is scattered by the cells. This type of sensor operates on the basis of binding with a known protein and is characterized as a bacterium onto a silicon chip, which can further bind with any other *E. coli* bacteria present in the food sample [105]. Chen and Durst [106] devised an array-based immunosorbent assay to detect *E. coli* O157:H7, *Salmonella* spp., and *L. monocytogenes* using protein G-liposomal nano-vesicles in pure and mixed cultures. The researchers showed that protein G-liposomal nanovesicles can be successfully used in immunoassays for simultaneous detection of foodborne pathogens and demonstrated the effectiveness of protein G-liposomal nano-vesicles as universal immunoassay reagents. Further, DeCory et al. [107] developed an immunomagnetic bead-immunoliposome fluorescence assay to rapidly detect *E. coli* O157:H7 in aqueous samples. The findings demonstrated the feasibility of using immunomagnetic beads in combination with sulforhodamine B encapsulated in immunoliposomes to rapidly detect *E. coli* O157:H7 in aqueous samples. Additionally, liposome-based methods have also been investigated by other researchers for pathogen detection [108], [109]. Nanosensors, such as nano-cantilevers, employ silicon-based materials to recognize proteins and detect pathogens that vibrate at different frequencies depending on their biomass [110].

In recent years, various new nanoparticle-based detection platforms have been developed. Tominaga [111] developed lateral-flow immune test strips with palladium nanoparticles against *Klebsiella*, which allowed for the specific binding and visualized detection of specific bacteria. Further, Thakur et al. [112] detected single *E. coli* bacterial cell using a reduced graphene, nanoparticle-based, field-effect transistor device. Recently, our research group developed an electrochemical sensing platform based on graphene oxide–gold nanoparticles to detect *Cronobacter* sakazakii, which is a bacterium that is hazardous to infants, in infant formula powder with a detection limit of 2.0×10^1 cfu/mL [113]. Also, Song et al. [114] developed a fluorescence sensing platform using immunomagnetic nanoparticles coupled with liposomenanoparticles to detect *Cronobacter* sp. at the genus level with a detection limit of 5.9×10^3 cfu/mL. Additionally, our research team developed an aptamer and gold nanoparticle-based optical detection platform to detect *Salmonella* in contaminated pork samples [115]. The various types of nanoparticles for the detection of various foodborne pathogens and their detection limits are shown in Table 2.

Table 2: Nanoparticles employed for the detection of foodborne pathogens

Nanoparticles	Pathogens	Detection limit	Reference
Gold nanoparticle	*Salmonella enterica serotype Typhi*	98.9 CFU/mL	[174]
Gold/silicon nanorod	*Salmonella enterica serotype Typhi*; Respiratory syncytial virus	Not reported	[175]
Gold nanorod	*Escherichia coli* O157:H7	1–10 CFU/mL	[159]
Quantum dot	*Salmonella enterica serotype Typhi, E. coli*O157:H7, *Listeria monocytogenes*	10^3–10^6 cells/mL	[160], [161], [162], [163]
Magnetic bead/ quantum dot	*E. coli* 0157:H7	10^3 CFU/mL	[176]
RuBpy doped silica	*E. coli* O157:H7	1 cell/mL	[164], [176]
Single walled carbon nanotube	*E. coli*	Not reported	[177]
Magnetic nanoparticle	*E. coli* O157:H7, *S. aureus, S. epidermidis*	10^4 CFU/mL, 8 CFU/mL, 10 CFU/mL	[177]
Immunomagnetic liposome nanoparticle	*Cronobacter sakazakii*	10^3 CFU/mL	[108]
Aptamer conjugated gold nanoparticles	*Salmonella typhimurium*	10^4 CFU/mL	[115], [120]
Liposome nanoparticles	*Salmonella typhimurium*	10^2 CFU/mL	[173]

Nanoparticles for Protection from Allergens

Nanotechnology has been employed as a basic tool to control and manage food allergens [116], [117]. Despite these efforts, certain nanomaterials cause allergic pulmonary inflammation [118], [119] in humans. For instance, SiO_2nanoparticles have been reported to cause allergen-specific, Th2-type immune responses *in vivo* in female BALB/c mice [118]. The immunotherapy of allergies using aluminum hydroxide (alum) as an adjuvant displayed several side effects, including swelling, indurations, erythema, cutaneous nodules, and granulomas at the injection site [119]. To overcome this, researchers have investigated the use of polymeric nanoparticles [120] and substances that target toll-like receptors (TLR) as alternative adjuvants [121]. For instance, the protamine-based nanoparticles with TLR-9 ligand cytosine phosphateguanine (CpG)-oligodeoxynucleotides (ODNs) can be used as an adjuvant. Protamines are arginine-rich peptides of approximately 4 kDa found in the sperm of salmon and are isolated from the mature testicles of salmons. Protamines are biodegradable and efficiently utilized for the reversal of heparin activity during surgical operations [122] or as an insulin-additive [123]. Additionally, protamine-based nanoparticles with CpG-ODN counteract the Th2-dominated immune response induced by an allergen, thus have shown remarkable potential in allergen immunotherapy as a novel carrier system [124].

Similarly, Gamazo et al. [125] reported immunotherapy that employed nanoparticles as an allergen delivery system, which facilitated the administration, reduced the dose, and diminished the allergen exposure to the immunoglobulin E that was bound to mast cells and/or basophils. Moreover, in recent years, bioinspired NSMs are being explored that are known to have little to no toxic effects and can thereby be potentially exploited in the food industry [126]. Localized surface plasmon resonance(LSPR)-based label-free biosensing methods have gained huge attention due to their high sensitivity, simplicity, and relatively low cost efficacy. However, the challenges have been observed using these methods for *in situ* analysis of real samples due to the instability of colloidal nanoparticles at certain levels of pH and salt concentration. Recently, Lee et al. [127] developed a simple and cost-effective optical fiber coated with aptamer-modified gold nanorods for easy detection of ochratoxin A in grape juice samples (food mycotoxin causing allergy) and subjected the fiber to LSPR analysis, which demonstrated a significant detection limit of 12.0 pM. Zhang et al. [128] reported a magnetic, nanoparticle-assisted, aptamer-based fluorescence assay to detect allergens in food matrices. As a result, under optimal conditions, the linear range was recorded as 0.4–5 g/mL ($R^2 = 0.996$), and a low limit of detection was calculated as 77 ng/mL with efficient selectivity. Moreover, Brotons-Canto et al. [129]evaluated the positive effects of poly-(anhydride) nanoparticles (150 nm) as oral vehicles for immunotherapy against experimental peanut allergies. Several other studies demonstrated the promising applications of nanotechnologies in vaccinology and in allergen immunotherapy [130]. In the near future, nanoparticle-based formulations can be applied to allergen immunotherapy to solve its unmet needs [131].

Nanoparticles for Preventing Heavy Metal Reduction

The release of heavy metals from nanomaterials poses a high risk of toxic outbreaks [80]. The release of these metals in food products has adverse effects in the case of long-term accumulation. Metal and metal oxide-based nanomaterials, such as ZnO [132], Ag [133], and CuO [134], increase the intracellular ROS level [132] and ultimately cause lipid peroxidation and DNA damage. Cationic surfactant coated silica-modified magnetite nanoparticlesact as adsorbents for the micro-extraction and have ability to determine trace amount of Cu, Ni, Co, Cd, Pb, and Mn from environmental samples. The silica-coated NPs synthesized using cetylpyridinium bromide, can solubilize metal ions after being complexed with 8-hydroxyquinoline [135]. While many nanomaterials have tremendous potential for the remediation of contaminants, magnetite nanoparticles have proven to be the most attractive and inexpensive substrates for the recovery of heavy metals [136] from various sources. Zhang et al. [137] fabricated $Fe@Fe_2O_3$ core/shells, nanowires, and nano-necklaces to remove chromium from aqueous solutions.

Recently, Lin et al. [138] demonstrated the ability of aminated magnetic iron oxide nanoparticles, especially Cu^{2+}, Ni^{2+}, Pb^{2+}, and Zn^{2+}, to act as adsorbents to remove aqueous heavy metal ions. The results of the study displayed that the increase in the degree of amination had a positive effect with a simultaneous increase in the adsorption capacity and the initial rate of adsorption of heavy metal ions. MgO nanoparticles synthesized via sol–geland calcination processes showed great potential against bacterial infections and were able to remove heavy metal ions from contaminated water samples [139]. These findings anticipated that nano-sized MgO particles may have potential for the treatment of bacterial and heavy metal contaminated wastewater due to their high removal efficiency, low cost, facile preparation, and environmentally friendly characteristics.

In recent years, Lingamdinne et al. [140] demonstrated how synthesized, reusable, recoverable iron oxide nanoparticles could reduce heavy metals without losing their stability. In addition, over the years, various carbon nanoparticles with high fluorescence properties such as carbon dots (C-dots) with a particle size of smaller than 10 nm and carbon nanoparticles (CNPs) of about 10 nm size or larger have shown a broad range of biological properties with low toxicity and good biocompatibility. Recently, through a thermal process in the presence of H_3PO_4, Simpson et al. [141] fabricated, characterized, and approved carbon nanoparticles (66 nm) with glycerol for the detection of heavy metal ions, with a detection limit as low as 0.30 ppm.

Nanoparticles for the Inhibition of Biofilm Formation

Biofilms are tightly-packed bunch of bacterial cells that adhere to many substrates and produce a polymeric extracellular matrix that is extremely difficult to penetrate [142]. The biofilm formation begins with the attachment of free floating microorganisms to the surface by employing van der Waals forces that create problems such as biocorrosion, biofouling, and accumulation in the food processing industries [143]. Glycerol monolaurate (GML), which is recognized as safe by the USFDA, acts as an antimicrobial agent against many Gram-positive bacteria, including *Bacillus anthracis* [144]. GML has been shown to inhibit the biofilm formation of three different strains of *S. aureus* and of MRSA [145]. Besides this, the use of antimicrobials throughout the nano-fibers of filter membranes is employed to prevent biofilm formation [137]. Moreover, Shahrokh and Emtiazi [146] demonstrated that nanosilver particles at low concentrations (0.2 ppm) enhanced bacterial metabolism and thus suggested the use of an optimum concentration of nanosilver particles for various nanomaterials to prevent biofilm formation.

Further, nickel oxide nanoparticles (NiO NPs) have been proposed as prospective antibacterial and anti-tumor agents. Through a green approach and using *Eucalyptus globulus* leaf extract, Saleem et al. [147] synthesized NiO-NPs that were 10–20 nm in size and assessed their anti-biofilm activity. In addition, Gambino et al. [148] demonstrated the bio-applicability of zinc oxide nanoparticles for hindering fungal biofilm (*Alternaria alternata*, *Penicillium chrysogenum*, *P. pinophilum,* and *Aspergillus niger*) that had developed in an ancient Egyptian tomb. Chlorhexidine-conjugated gold nanoparticlesexhibited biofilm inhibitory effects against *Klebsiella pneumonia* [149].

Bacillus subtilis has shown enormous potential in various industrial processes for the production of value-added products. Fermentation process of *B. subtilis*is directly related to its ability to form biofilm resulting in operational issues and major process. Ranmadugala et al. [150] evaluated the effects of two types of superparamagnetic iron oxide (IONs), such as naked IONs and IONs coated with 3-aminopropyltriethoxy silane (IONs@APTES) against *B. subtilis* in terms of their ability to inhibit its biofilm formation, bacterial growth, and cell viabilities. The results showed that a significant reduction in the total biomass of the bacterial biofilm was observed without affecting the cell viabilities. These results suggested that these nanoparticles could be applied in several industries to combat the growth of bacterial biofilm. Moreover, Thuptimdang et al. [151] suggested that controlling the growth conditions to alter the biofilm physical structure was a possible approach to reduce the impact of silver nanoparticles (AgNPs) on biofilms in engineered and natural systems. These recent findings are increasing the demand for researchers to explore the use of nanoparticles in several other industries, including medicine.

ARE NANOTECHNOLOGY AND BIG DATA EFFECTIVE ENOUGH FOR NEXT INDUSTRIAL REVOLUTION FOR SECURING "SMART FOOD"?

The engineering applications of nanotechnology will herald a new digital future characterized by improvements in food safety, shelf-life performance, and food reliability. These advancements are frequently referred to as big data information and open new fields of research and applications that will increasingly affect all parts of society. There is a huge complicated response to outbreaks associated with foodborne diseases due to the globalization of food supply chains. Generation of big data in the food industry may help to facilitate encouraging advancement in food safety and security at the global scale [156]. Big data is represented as high volume, high velocity, high veracity, and/or high variety information assets requiring new processing formats so as to enhance the process of decision making, process optimization, and insight discovery [157]. Although the big data analysis and large datasets in food safety and quality have focused on improving root cause and retrospective analyses, there is a huge growth in the advancement and utilization of predictive analytics in food security sector [157]. For example, whole genome sequencing data for *L. monocytogenes* isolates obtained from ice cream in Kansas became publicly available soon after a listeriosis outbreak linked to ice cream was reported in early 2015. Other omics datasets, such as metagenomics data, have also been used to identify and characterize food spoilage issues. It is likely that these types of data sources will become increasingly available to personnel in the food industry.

Recently, the World Health Organization (WHO) has enfold the big data information approach to deal with the food-safety process of decision making, which has brought about a new food-safety platform known as "FOSCOLLAB," which integrates different sources from various disciplines [158]. On this platform, organized and nano-structured information from numerous areas, for example, food, horticulture, animals, public health, and economic index, are incorporated and accessible to the users by means of a few committed dashboards [158]. The information sources associated in FOSCOLLAB are assessments of the Joint FAO/WHO Expert Committee on Food Additives and the Joint FAO/WHO Meeting on Pesticide Residues databases on chemical risk assessment, the WHO database on Collaborating Centers, and the Global Environment Monitoring System (GEMS) food databases on chemical occurrence and food consumption. There is no restriction of utilizing big data information in the fields of food safety and food quality to facilitate improved methodologies.

Finally, input food industries often work worldwide, are R&D intensive, and deal with new technologies, including genetics, robotics, informatics, and nanotechnology. Besides the challenge of feeding the world, there is the issue of sustainability, which is mainly an issue of farming and transport. Together with the relationship between food and health (e.g. obesity concerns), food safety directly affects food processors, retailers, and consumers. The ultimate objective is to estimate risks to farms and to build up a preventive way to deal with food safety issues. The use of geographical information system technology (GIST) to create food security has demonstrated incredible practices, for example, helping producers settle on more informed choices about field practices and establish the programs of targeted pathogen surveillances. Overall, the utilization of GIST to yield big data information in food safety investigations has created remarkable measures of new data with respect to the ecology of various life forms in the food environment and data on different contamination episodes. In the era of big data information, GIST is one of the data tools that enables scientists to store, analyze, process, capture, and visualize extensive datasets. Additionally, the use of GIST to complex food safety/security issues is being demonstrated for further integration of various large data sets. Essential components associated with big data requirements in nano-food safety are presented in Fig. 3.

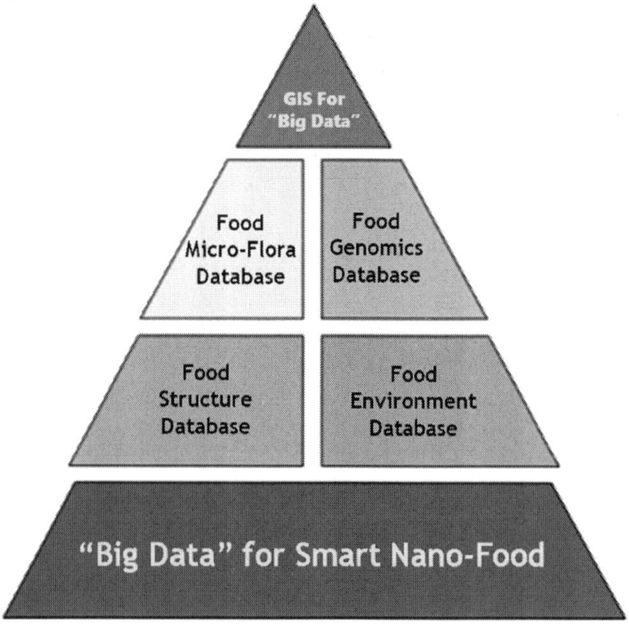

Figure 3: Essential components for producing "big data" for safety and security of smart nano-food.

FUTURE PERSPECTIVES AND POTENTIAL RISKS OF NANOTECHNOLOGY

There have been tremendous developments in the application of nanotechnology in food science and research. Nanotechnology helps in the detection of pesticides [159], pathogens [160], and toxins [161], and tracking-tracing-monitoring helps in the maintenance of food quality. A massive amount of interest is being on using carbon nano-tubes for incorporating them into packaging materials for the detection of toxic proteins, microorganisms, and food spoilage [162]. Furthermore, carbon nanotube can transform our future food packaging materials to active and intelligent packaging systems [163]. Along with the incessant research on the application of nanomaterials, several potential risks and toxicity issues associated with the application of nanomaterials have been revealed, and these concerns must be addressed [104], [161]. The effect of these nanoscale particles on human beings, animals, and the environment are unpredictable due to changes over time in their properties. Some nanoparticles can even cross biological barriers, such as the blood–brain barrier, and enter various cells and organs [164].

The evidences for the health risks associated with the inhalation of ultrafine and nanoparticles have been increasing every day [60]. In addition, bioaccumulation of nanomaterials, such as nano-silver, derived from either nanopackaging or plants and animals has been confirmed in food and human being [62]. The use of nanomaterials as activation catalysts, pesticides, and microbicides always pose an unknown associated risk, and hence risk assessment procedures must be strictly followed while processing food [57], [58]. Even with the advent of nanotechnology, the challenges to the development of a healthy and sustainable food industry remain. With the introduction and development of nanotechnology in the food system, the public should be educated regarding the associated health, safety, and environmental effects of nanotechnology.

ACKNOWLEDGEMENTS

We acknowledge financial support from the Basic Science Research Program through the National Research Foundation of Korea (NRF) funded by the Ministry of Science, ICT & Future Planning (2016R1A2B4013374 and 2014R1A5A1009799).

REFERENCES

1. Chandra P. Nanobiosensors for personalized and onsite biomedical diagnosis. The Institution of Engineering and Technology; 2016. https://doi.org/10.1049/PBHE001E.
2. Sandoval B. Perspectives on FDA's regulation of nanotechnology: emerging challenges and potential solutions. Compr Rev Food Sci Food Saf 2009;8:375e93.
3. Moriarty P. Nanostructured materials. Rep Prog Phys 2001;64:297.
4. Pathkoti K, Manubolu M, Hwang HM. Nanostructures: current uses and future applications in food science. J Food Drug Anal 2017;25(2):245e53.
5. Bhushan B. Springer handbook of nanotechnology. Berlin; Heidelberg: Springer Science & Business Media; 2010.
6. Gupta A, Eral HB, Hatton TA, Doyle PS. Nanoemulsions: formation, properties and applications. Soft Matter 2016;12:2826e41.
7. Singh T, Shukla S, Kumar P, Wahla V, Bajpai VK, Rather IA. Application of nanotechnology in food Science: perception and overview. Front Microbiol 2017;8:e1501.
8. Fu PP. Introduction to the special issue: nanomaterials e toxicology and medical applications. J Food Drug Anal 2014;22:1e2.
9. Hu B, Liu X, Zhang C, Zeng X. Food macromolecule based nanodelivery systems for enhancing the bioavailability of polyphenols. J Food Drug Anal 2017;25:3e15.
10. Baranwal A, Srivastava A, Kumar P, Bajpai VK, Maurya PK, Chandra P. Prospects of nanostructure materials and their composites as antimicrobial agents. Front Microbiol 2018;9:e422.
11. Das MP, Livingstone JR, Veluswamy P, Das J. Exploration of Wedelia chinensis leaf-assisted silver nanoparticles for antioxidant, antibacterial and in vitro cytotoxic applications. J Food Drug Anal 2017;26:917e25.
12. He X, Hwang HM. Nanotechnology in food science: functionality, applicability, and safety assessment. J Food Drug Anal 2016;24:671e81.
13. Adabi M, Naghibzadeh M, Adabi M, Ali Zarrinfard M, Esnaashari SS, Seifalian AM, et al. Biocompatibility and nanostructured materials: applications in nanomedicine. Artif Cells Nanomed Biotechnol 2017;45:833e42.

14. He X, Aker WG, Pelaez M, Lin Y, Dionysiou DD, Hwang HM. Assessment of nitrogenefluorine-codoped TiO2 under visible light for degradation of BPA: implication for field remediation. J Photochem Photobiol Chem 2016;314:81e92.

15. Wang B, Feng WY, Wang TC, Jia G, Wang M, Shi JW, et al. Acute toxicity of nano- and micro-scale zinc powder in healthy adult mice. Toxicol Lett 2006;16:115e23.

16. He X, Aker WG, Hwang HM. An in vivo study on the photoenhanced toxicities of S-doped TiO2 nanoparticles to zebra fish embryos (Danio rerio) in terms of malformation, mortality, rheotaxis dysfunction, and DNA damage. Nanotoxicology 2014;8(S1):185e95.

17. Peters RJ, Bouwmeester H, Gottardo S, Amenta V, Arena M, Brandhoff P. Nanomaterials for products and application in agriculture, feed and food. Trends Food Sci Technol 2016;54:155e64.

18. Nanotechnology information statement. IFST e Institute of Food Science & Technology; 2006.

19. Robertson GL. Food packaging: principles and practice. 2nd ed. New York: CRC Press, Taylor & Francis Group; 2006. ISBN 9781439862414.

20. Finnigan B. Barrier polymers. In: Yam KL, editor. The wiley encyclopedia of packaging technology. New York: John Wiley and Sons Inc; 2009. p. 103e9.

21. Abbaspour A, Norouz-Sarvestani F, Noori A, Soltani N. Aptamer-conjugated silver nanoparticles for electrochemical dual-aptamer-based sandwich detection of Staphylococcus aureus. Biosens Bioelectron 2015;68:149e55.

22. Ghanbarzadeh B, Oleyaei SA, Almasi H. Nano-structured materials utilized in biopolymer based plastics for food packaging applications. Crit Rev Food Sci Nutr 2014;7:37e41.

23. Abdollahi M, Rezaei M, Farzi G. A novel active bionanocomposite film incorporating rosemary essential oil and nanoclay into chitosan. J Food Eng 2012;111(2):343e50.

24. Azeredo HMC, Mattoso LHC, McHugh TH. Nanocomposites in food packagingda review. In: Reddy B, editor. Advances in diverse industrial applications of nanocomposites. InTech; 2011. p. 550. https://doi.org/10.5772/1931.

25. Klaine SJ, Koelmans AA, Horne N, Carley S, Handy RD, Kapustka L. Paradigms to assess the environmental impact of manufactured nanomaterials. Environ Toxicol Chem 2012;31(1):3e14.
26. Klaine SJ. Considerations for research on the environmental fate and effects of nanoparticles. Environ Toxicol 2009;28(9):1787e8.
27. Hannon JC, Kerry J, Cruz-Romero M, Morris M, Cummins E. Advances and challenges for the use of engineered nanoparticles in food contact materials. Trends Food Sci Technol 2015;43:43e62.
28. Echegoyen Y, Nerin C. Nanoparticle release from nanosilver antimicrobial food containers. Food Chem Toxicol 2013;62:16e22.
29. Mills A. Oxygen indicators and intelligent inks for packaging food. Chem Soc Rev 2005;34:1003e11.
30. Arora A, Padua GW. Review: nanocomposites in food packaging. J Food Sci 2010;75:43e9.
31. Stormer A, Bott J, Kemmer D, Franz R. Critical review of the migration potential of nanoparticles in food contact plastics. Trends Food Sci Technol 2017;63:39e50.
32. Arndt GW. Leak testing. In: Yam KL, editor. The wiley encyclopedia of packaging technology. 3rd ed. New York: John Wiley and Sons, Inc.; 2008.
33. Duncan TV. Applications of nanotechnology in food packaging and food safety: barrier materials, antimicrobials and sensors. J Colloid Interface Sci 2011 2011;363:1e24.
34. Ariyarathna IR, Rajakaruna RMPI, Nedra Karunaratne D. The rise of inorganic nanomaterial implementation in food applications. Food Control 2017;77:251e9.
35. Yu Z, Lib B, Chud J, Zhang P. Silica in situ enhanced PVA/chitosan biodegradable films for food packages. Carbohydr Polym 2017;184:214e20.
36. Swaroop C, Shukla M. Nano-magnesium oxide reinforced polylactic acid biofilms for food packaging applications. Int J Biol Macromol 2018;113:729e36.
37. Foltynowicz Z, Bardenshtein A, Sangerlaub S, Antvorskov H, € Kozak W. Nanoscale, zero valent iron particles for application as oxygen scavenger in food packaging. Food Packag Shelf Life 2017;11:74e83.

38. Dasgupta N, Ranjan S, Mundekkad D, Ramalingam C, Shanker R, Kumar A. Nanotechnology in agrofood: from field to plate. Food Res Int 2015;69:381e400.

39. Silvestre C, Duraccio D, Cimmino S. Food packaging based on polymer nanomaterials. Prog Polym Sci 2011;36:1766e82.

40. Ghaani M, Cozzolino CA, Castelli G, Farris S. An overview of the intelligent packaging technologies in the food sector. Trends Food Sci Technol 2016;51:1e11.

41. Fang Z, Zhao Y, Warner RD, Johnson SK. Active and intelligent packaging in meat industry. Trends Food Sci Technol 2017;61:60e71.

42. Silva HD, Cerqueira MA, Vicente AA. Nanoemulsions for food applications: development and characterization. Food Bioprocess Technol 2012;5(3):854e67.

43. Tamjidi F, Nasirpour A, Shahedi M. Physicochemical and sensory properties of yogurt enriched with microencapsulated fish oil. Food Sci Technol Int 2012;18(4):381e90.

44. De Conto LC, Grosso CRF, Gonc¸alves LAG. Chemometry as applied to the production of omega-3 microcapsules by complex coacervation with soy protein isolate and gum Arabic. LWT e Food Sci Technol 2013;53(1):218e24.

45. Li XY, Chen XG, Sun ZW, Park HJ, Cha DS. Preparation of alginate/chitosan/carboxymethyl chitosan complex microcapsules and application in Lactobacillus casei ATCC 393. Carbohydr Polym 2011;83(4):1479e85.

46. Zuidam NJ, Shimoni E. Overview of microencapsulates for use in food products or processes and methods to make them. In: Encapsulation technologies for active food ingredients and food processing. Springer Science Business Media LLC; 2010.

47. Gibbs BF, Kermash S, Alli I, Mulligan CN. Encapsulation in the food industry: a review. Int J Food Sci Nutr 2010;199:213e24.

48. Murugesan R, Orsat V. Spray drying for the production of nutraceutical ingredients e a review. Food Bioprocess Technol 2012;5(1):3e14.

49. Dekkers S, Krystek P, Peters RJ, Lankveld DX, Bokkers BG, van Hoeven-Arentzen PH, et al. Presence and risks of nanosilica in food products. Nanotoxicology 2011;5:393e405.

50. Mozafari MR, Flanagan J, Matia-Merino L, Singh H. Recent trends in the lipid-based nanoencapsulation of antioxidants and their role in food. J Sci Food Agric 2006;86(13):2038e45.

51. Ezhilarasi PN, Karthik P, Chhanwal N, Anandharamakrishnan C. Nanoencapsulation techniques for food bioactive components: a review. Food Bioprocess Technol 2013;6:628e47.

52. Flores-Lopez ML, Cerqueira MA, de Rodrıguez DJ, Vicente AA. Perspectives on utilization of edible coatings and nano-laminate coatings for extension of postharvest storage of fruits and vegetables. Food Eng Rev 2015;5:1e14.

53. Fakhouri FM, Casari ACA, Mariano M, Yamashita F, Mei LI, Soldi V, et al. Effect of a gelatin-based edible coating containing cellulose nanocrystals (CNC) on the quality and nutrient retention of fresh strawberries during storage, IOP conference series: materials science and engineering, vol. 64, Conference 1, 2nd International Conference on Structural Nano Composites (NANOSTRUC 2014), 2014; 20e21. May 2014. Madrid, Spain.

54. Shi S, Wang W, Liu L, Wu S, Wei Y, Li W. Effect of chitosan/ nano-silica coating on the physicochemical characteristics of longan fruit under ambient temperature. J Food Eng 2013;118:125e31.

55. Yu YW, Zhang SY, Ren YZ, Li H, Zhang XN, Di JH. Jujube preservation using chitosan film with nano-silicon dioxide. J Food Eng 2012;113:408e14.

56. Medeiros BS, Souza MP, Pinheiro AC, Bourbon AI, Cerqueira MA, Vicente AA, et al. Physical characterization of an alginate/lysozyme nano-laminate coating and its evaluation on 'Coalho' cheese shelf life. Food Bioprocess Technol 2014;7:1088e98.

57. Yang FM, Li HM, Li F, Xin ZH, Zhao LY, Zheng YH, et al. Effect of nano-packaging on preservation quality of fresh strawberry (Fragaria ananassa Duch. cv Fengxiang) during storage at 4 C. J Food Sci 2010;75(3):236e40.

58. Elgadir MA, Uddin MS, Ferdosh S, Adam A, Khan AJ, Saraker MZI. Impact of chitosan composites and chitosan nanoparticle composites on various drug delivery systems: a review. J Food Drug Anal 2015;23(4):619e29.

59. Jovanovic B. Critical review of public health regulations of titanium dioxide, a human food additive. Integr Environ Assess Manag 2015;11:10e20.

60. Johnston CT. Probing the nanoscale architecture of clay minerals. Clay Miner 2010;45(3):245e79.

61. Fujishima A, Rao TN, Tryk DA. Titanium dioxide photocatalysis. J Photochem Photobiol C Photochem Rev 2000;1:1e21.

62. Chawengkijwanich C, Hayata Y. Development of TiO2 powder-coated food packaging film and its ability to inactivate Escherichia coli in vitro and in actual tests. Int J Food Microbiol 2008;123:288e92.

63. Qi LF, Xu ZR, Jiang X, Hu C, Zou X. Preparation and antibacterial activity of chitosan nanoparticles. Carbohydr Res 2004;339:2693e700.

64. Sarwar MS, Niazi MBK, Jahan Z, Ahmad T, Hussain A. Preparation and characterization of PVA/nanocellulose/Ag nanocomposite films for antimicrobial food packaging. Carbohydr Polym 2018;184:453e64.

65. Valerinia D, Tammaroa L, Di Benedettoa F, Vigliottab G, Capodiecia L, Terzia R, et al. Aluminum-doped zinc oxide coatings on polylactic acid films for antimicrobial food packaging. Thin Solid Films 2018;645:187e92.

66. Lu WC, Huang DW, Wang CCR, Yeh CH, Tsai JC, Huang YT, et al. Preparation, characterization, and antimicrobial activity of nanoemulsions incorporating citral essential oil. J Food Drug Anal 2018;26:82e9.

67. Liau SY, Read DC, Pugh WJ, Furr JR, Russell AD. Interaction of silver nitrate with readily identifiable groups: relationship to the antibacterial action of silver ions. Lett Appl Microbiol 1997;25:279e83.

68. Sondi I, Salopek-Sondi B. Silver nanoparticles as antimicrobial agent: a case study on E. coli as a model for Gram-negative bacteria. J Colloid Interface Sci 2004;275:177e82.

69. Li Y, Tseng YD, Kwon SY, Despaux L, Bunch JS, Mceuen PL. Controlled assembly of dendrimer-like DNA. Nat Mater 2004;3:38e42.

70. Morones JR, Elechiguerra JL, Camacho A, Holt K, Kouri JB, Ramirez JT. The bactericidal effect of silver nanoparticles. Nanotechnology 2005;16:2346e53.

71. Karimi M, Sadeghi R, Kokini J. Human exposure to nanoparticles through trophic transfer and the biosafety concerns that nanoparticle-contaminated foods pose to consumers. Trends Food Sci Technol 2018;75:129e45.

72. Suppakul P, Miltz J, Sonneveld K, Bigger SW. Active packaging technologies with an emphasis on antimicrobial packaging and its

applications. J Food Sci 2003;68:408e20.

73. Ahmed I, Lin H, Zou L, Brody AL, Li Z, Qazi IM, et al. A comprehensive review on the application of active packaging technologies to muscle foods. Food Control 2017;82:163e78.
74. Arfat YA, Ahmed J, Hiremath N, Auras R, Joseph A. Thermomechanical, rheological, structural and antimicrobial properties of bionanocomposite films based on fish skin gelatin and silver-copper nanoparticles. Food Hydrocolloids 2017;62:191e202.
75. Manke A, Wang L, Rojanasakul Y. Mechanisms of nanoparticle-induced oxidative stress and toxicity. BioMed Res Int 2013;15:e942916.
76. Pathakoti K, Morrow S, Han C, Pelaez M, He X, Dionysiou DD, et al. Photo-inactivation of Escherichia coli by sulfur and nitrogen-fluorine-codoped TiO2 nanoparticles under solar simulated light and visible light irradiation. Environ Sci Technol 2013;47:9988e96.
77. He X, Aker WG, Huang MJ, Watts JD, Hwang HM. Metal oxide nanomaterials in nanomedicine: applications in photodynamic therapy and potential toxicity. Curr Top Med Chem 2015;15:1887e900.
78. Pool H, Quintanar D, de Dios Figueroa J, Mano CM, Bechara JEH, Godinez LA, et al. Antioxidant effects of quercetin and catechin encapsulated into PLGA nanoparticles. J Nanomater 2012;2012:e145380.
79. Rojas-Grau MA, Soliva-Fortuny R, Martin-Belloso O. Edible coatings to incorporate active ingredients to fresh cut fruits: a review. Trends Food Sci Technol 2009;20:438e47.
80. Li M, Zhu L, Lin D. Toxicity of ZnO nanoparticles to Escherichia coli: mechanism and the influence of medium components. Environ Sci Technol 2011;45(5):1977e83.
81. Inbaraj BS, Chen BH. Nanomaterial-based sensors for detection of foodborne bacterial pathogens and toxins as well as pork adulteration in meat products. J Food Drug Anal 2016;24:15e28.
82. Momin JK, Jayakumar C, Prajapati JB. Potential of nanotechnology in functional foods. Emir J Food Agric 2013;25:10e9.
83. Liang J, Yan H, Wang X, Zhou Y, Gao X, Puligundla P, et al. Encapsulation of epigallocatechin gallate in zein/chitosan nanoparticles for controlled applications in food systems. Food Chem 2017;231:19e24.
84. Youssef AM. Polymer nanocomposites as a new trend for packaging applications. Polym Plast Technol Eng 2013;52:635e60.

85. Laoutid F, Bonnaud L, Alexandre M, Lopez-Cuesta JM, Dubois P. New prospects in flame retardant polymer materials: from fundamentals to nanocomposites. Mater Sci Eng R Rep 2009;63:100e25.

86. Lizundia E, Ruiz-Rubio L, Vilas JL, Leon LM. Poly (l-lactide)/ ZnO nanocomposites as efficient UV-shielding coatings for packaging applications. J Appl Polym Sci 2016;133:e42426.

87. USFDA. Listing of color additives exempt from certification. In: Office USGP, editor. Code of federal regulations title 21dfood and drugs. Washington, DC: U.S. FDA; 2002

88. Accessed 18.07.16..

89. USFDA. Color additive status list. 2015. http://www.fda.gov/ForIndustry/ColorAdditives/ColorAdditiveInventories/ ucm106626.html Accessed 18.07.16..

90. Mehrad B, Ravanfar R, Licker J, Regenstein JM, Abbaspourrad A. Enhancing the physicochemical stability of b-carotene solid lipid nanoparticle (SLNP) using whey protein isolate. Food Res Int 2018;105:962e9.

91. Pal M. Nanotechnology: a new approach in food packaging. J Food e Microbiol Safety Hyg 2017;2:121.

92. Wesley SJ, Raja P, Sunder Raj AA, Tiroutchelvamae D. Review on nanotechnology applications in food packaging and safety. Int J Eng Res 2014;3:645e51.

93. Chandra P, Noh HB, Won MS, Shim YB. Detection of daunomycin using phosphatidylserine and aptamer coimmobilized on Au nanoparticles deposited conducting polymer. Biosens Bioelectron 2011;26(11):4442e9.

94. Thakur MS, Ragavan KV. Biosensors in food processing. J Food Sci Technol 2013;50(4):625e41.

95. Li YS, Church JS. Raman spectroscopy in the analysis of food and pharmaceutical nanomaterials. J Food Drug Anal 2014;22(1):28e48.

96. Jarvis RM, Goodacre R. Discrimination of bacteria using surface enhanced Raman spectroscopy. Anal Chem 2004;76:40e7.

97. Kahraman M, Yazıcı MM, Sahin F, Culha M. Convective assembly of bacteria for surface-enhanced Raman scattering. Langmuir 2008;24:894e901.

98. Baranwal A, Mahato K, Srivastava A, Maurya PK, Chandra P. Phytofabricated metallic nanoparticles and their clinical applications.

RSC Adv 2016;6(107):105996e6010.

99. Zuo P, Li X, Dominguez DC, Ye BC. A PDMS/paper/glass hybrid microfluidic biochip integrated with aptamerfunctionalized graphene oxide nano-biosensors for onestep multiplexed pathogen detection. Lab Chip 2013;13:3921e8.

100. Holzinger M, Le Goff A, Cosnier S. Nanomaterials for bio sensing applications: a review. Front Chem 2014;2:e63.

101. Yang M, Peng Z, Ning Y, Chen Y, Zhou Q, Deng L. Highly specific and cost-efficient detection of Salmonella paratyphi combining aptamers with single-walled carbon nanotubes. Sensors 2013;13:6865e81.

102. Bhattacharya S, Jang J, Yang L, Akin D, Bashir R. Biomems and nanotechnology based approaches for rapid detection of biological entities. J Rapid Meth Autom Microbiol 2007;15:1e32.

103. Chen CS, Durst RA. Simultaneous detection of Escherichia coli O157:H7, Salmonella spp. and Listeria monocytogenes with an array-based immunosorbent assay using universal protein G-liposomal nanovesicles. Talanta 2006;69:232e8.

104. DeCory TR, Durst RA, Zimmerman SJ, Garringer LA, Paluca G, DeCory HH, et al. Development of an immunomagnetic bead immunoliposome fluorescence assay for rapid detection of Escherichia coli O157:H7 in aqueous samples and comparison of the assay with a standard microbiological method. Appl Environ Microbiol 2005;71:1856e64.

105. Shukla S, Lee G, Song X, Park S, Kim M. Immunoliposomebased immunomagnetic concentration and separation assay for rapid detection of Cronobacter sakazakii. Biosens Bioelectron 2016;77:986e94.

106. Shukla S, Leem H, Kim M. Development of a liposomebased immunochromatographic strip assay for the detection of Salmonella. Anal Bioanal Chem 2011;401:2581e90.

107. Jain KK. Nanodiagnostics: application of nanotechnology in molecular diagnostics. Expert Rev Mol Diagn 2003;3(2):153e61.

108. Tominaga T. Rapid detection of Klebsiella pneumoniae, Klebsiella oxytoca, Raoultella ornithinolytica and other related bacteria in food by lateral-flow test strip immunoassays. J Microbiol Meth 2018;147:43e9.

109. Thakur B, Zhou G, Chang J, Pu H, Jin B, Sui X, et al. Rapid detection of single E. coli bacteria using a graphene-based field-effect transistor device. Biosens Bioelectron 2018. https://doi.org/10.1016/j.bios.2018.03.014.

110. Shukla S, Haldorai Y, Bajpai VK, Rengaraj A, Hwang SK, Song X, et al. Electrochemical coupled immunosensing platform based on graphene oxide/gold nanocomposite for sensitive detection of Cronobacter sakazakii in powdered infant formula. Biosens Bioelectron 2018;30:139e49.

111. Song X, Shukla S, Kim M. Detection of Cronobacter species in powdered infant formula using immunoliposome-based immunomagnetic concentration and separation assay. Food Microbiol 2018;72:23e30.

112. Oh SY, Heo NS, Shukla S, Cho HY, Ezhil Vilian AT, Kim J, et al. Development of gold nanoparticle aptamer-based LSPR sensing chips for the rapid detection of Salmonella typhimurium in pork meat. Sci Rep 2017;7:e10130.

113. Kumar S, Dilbaghi N, Barnela M, Bhanjana G, Kumar R. Biosensors as novel platforms for detection of food pathogens and allergens. BioNano Sci 2012;2:196e217.

114. Pilolli R, Monaci L, Visconti A. Advances in biosensor development based on integrating nanotechnology and applied to food-allergen management. Trends Anal Chem 2013;47:12e26.

115. Yoshida T, Yoshioka Y, Fujimura M, Yamashita K, Higashisaka K, Morishita Y, et al. Promotion of allergic immune responses by intranasal-administrated nanosilica particles in mice. Nanoscale Res Lett 2011;6:1e6.

116. Vogelbruch M, Nuss B, Korner M, Kapp A, Kiehl P, Bohm W. Aluminium induced granulomas after inaccurate intradermal hyposensitization injections of aluminiumadsorbed depot preparations. Allergy 2000;55:883e7.

117. De Souza Reboucas J, Esparza I, Ferrer M, Sanz ML, Irache JM, Gamazo C. Nanoparticulate adjuvants and delivery systems for allergen immunotherapy. J Biomed Biotechnol 2012:e474605.

118. Hedayat M, Takeda K, Rezaei N. Prophylactic and therapeutic implications of toll-like receptor ligands. Med Res Rev 2012;32:294e325.

119. Schulman S, Bijsterveld NR. Anticoagulants and their reversal. Transfus Med Rev 2007;21:37e48.

120. Owens DR. Insulin preparations with prolonged effect. Diabetes Technol Therapeut 2011;13(S1):S5e14.

121. Pali-Scholl I, Szollosi H, Starkl P, Scheicher B, Stremnitzer C, Hofmeister A. Protamine nanoparticles with CpGoligodeoxynucleotide prevent an allergen-induced Th2- response in BALB/c mice. Eur J Pharm Biopharm 2013;85:656e64.

122. Gamazo C, Gastaminza G, Ferrer M, Sanz ML, Irache JM. Nanoparticle based-immunotherapy against allergy. Immunotherapy 2014;6(7):885e97.

123. Prasad A, Mahato K, Chandra P, Srivastava A, Joshi SN, Maurya PK. Bioinspired composite materials: applications in diagnostics and therapeutics. J Mol Eng Mater 2016;4:e1640004.

124. Lee B, Park JH, Byun JY, Kim JH, Kim MG. An optical fiberbased LSPR aptasensor for simple and rapid in-situ detection of ochratoxin A. Biosens Bioelectron 2018;102:504e9.

125. Zhang Y, Wu Q, Sun M, Zhang J, Mo S, Wang J, et al. Magnetic-assisted aptamer-based fluorescent assay for allergen detection in food matrix. Sensor Actuator B Chem 2018;263:43e9.

126. Brotons-Cantoa A, Gamazo C, Martı´n-Arbella N, Abdulkarim M, Matı´as J, Gumbleton M, et al. Evaluation of nanoparticles as oral vehicles for immunotherapy against experimental peanut allergy. Int J Biol Macromol 2018;110:328e35.

127. Di Felice G, Barletta B, Bonura A, Butteroni C, Corinti S, Colombo P. Nanoparticles adjuvants in allergology: new challenges and pitfalls. Curr Pharmaceut Des 2015;21:4229e39.

128. Di Felice G, Colombo P. Nanoparticleeallergen complexes for allergen immunotherapy. Int J Nanomed 2017;12:4493e504.

129. Fukui H, Horie M, Endoh S, Kato H, Fujita K, Nishio K. Association of zinc ion release and oxidative stress induced by intratracheal instillation of ZnO nanoparticles to rat lung. Chem Biol Interact 2012;198:29e37.

130. McShan D, Ray PC, Yu H. Molecular toxicity mechanism of nanosilver. J Food Drug Anal 2014;22:116e27.

131. Karlsson HL, Cronholm P, Hedberg Y, Tornberg M, Battice LD, Svedhem S, et al. Cell membrane damage and protein interaction induced by copper containing nanoparticles-importance of the metal release process. Toxicology 2013;313:59e69.

132. Karatapanis AE, Fiamegos Y, Stalikas CD. Silica-modified magnetic nanoparticles functionalized with cetylpyridinium bromide for the preconcentration of metals after complexation with 8-hydroxy-

quinoline. Talanta 2011;84:834e9.

133. Amin MT, Alazba AA, Manzoor U. A review of removal of pollutants from water/wastewater using different types of nanomaterials. Adv Mater Sci Eng 2014:1e24.

134. Zhang L, Luo J, Menkhaus TJ, Varadaraju H, Sun Y, Fong H. Antimicrobial nano-fibrous membranes developed from electrospun polyacrylonitrile nanofibers. J Membr Sci 2011;369:499e505.

135. Lin S, Xu M, Zhang W, Hu X, Lin K. Quantitative effects of amination degree on the magnetic iron oxide nanoparticles (MIONPs) using as adsorbents to remove aqueous heavy metal ions. J Hazard Mater 2017;335:47e55.

136. Cai Y, Li C, Wu D, Wang W, Tan F, Wang X, et al. Highly active MgO nanoparticles for simultaneous bacterial inactivation and heavy metal removal from aqueous solution. Chem Eng J 2017;312:158e66.

137. Lingamdinne LP, Chang YY, Yang JK, Singh J, Choi EH, Shiratani M, et al. Biogenic reductive preparation of magnetic inverse spinel iron oxide nanoparticles for the adsorption removal of heavy metals. Chem Eng J 2017;307:74e84.

138. Simpson A, Pandey RR, Chusuei CC, Ghosh K, Patel R, Wanekaya AK. Fabrication characterization and potential applications of carbon nanoparticles in the detection of heavy metal ions in aqueous media. Carbon 2018;127:122e30.

139. Forier K, Raemdonck K, De Smedt SC, Demeester J, Coenye T, Braeckmans K. Lipid and polymer nanoparticles for drug delivery to bacterial biofilms. J Control Release 2014;190:607e23.

140. Shakeri S, Kermanshahi R, Moghaddam M, Emtiazi G. Assessment of biofilm cell removal and killing and biocide efficacy using the microtiter plate test. Biofouling 2007;23:79e86.

141. Vetter SM, Schlievert PM. Glycerol monolaurate inhibits virulence factor production in Bacillus anthracis. Antimicrob Agents Chemother 2005;49:1302e5.

142. Schlievert PM, Peterson ML. Glycerol monolaurate antibacterial activity in broth and biofilm cultures. PLoS One 2012;7:e40350.

143. Shahrokh S, Emtiazi G. Toxicity and unusual biological behavior of nanosilver on Gram positive and negative bacteria assayed by microtiter-plate. Eur J Biol Sci 2009;1:28e31.

144. Saleem S, Ahmed B, Khan MS, Al-Shaeri S, Musarrat J. Inhibition

of growth and biofilm formation of clinical bacterial isolates by NiO nanoparticles synthesized from Eucalyptus globulus plants. Microb Pathog 2017;111:375e87.

145. Gambino M, Ali Ahmed MA, Villa F, Cappitelli F. Zinc oxide nanoparticles hinder fungal biofilm development in an ancient Egyptian tomb. Int J Biodeterior Biodegrad 2017;122:92e9.

146. Ahmed A, Khan AK, Anwar A, Ali SA, Shah MR. Biofilm inhibitory effect of chlorhexidine conjugated gold nanoparticles against Klebsiella pneumonia. Microb Pathog 2016;98:50e6.

147. Ranmadugala D, Ebrahiminezhad A, Manley-Harris M, Ghasemid Y, Berenjian A. The effect of iron oxide nanoparticles on Bacillus subtilis biofilm, growth and viability. Process Biochem 2017;62:231e40.

148. Thuptimdang P, Limpiyakorn T, Khan E. Dependence of toxicity of silver nanoparticles on Pseudomonas putida biofilm structure. Chemosphere 2017;188:199e207.

149. King T, Cole M, Farber JM, Eisenbrand G, Zabaras D, Fox EM, et al. Food safety for food security: relationship between global megatrends and developments in food safety. Trends Food Sci Technol 2017;68:160e75.

150. Wiedmann M. Can big data revolutionize food safety? Food quality and safety. 2015. Retrieved from: http://www. foodqualityandsafety. com/article/can-bigdatarevolutionize-food-safety/.

151. WHO e World Health Organization, FOSCOLLAB. Global platform for food safety data and information. 2015. http://www.who.int/ foodsafety/foscollab_dashboards/ en/.

152. Fu J, Park B, Siragusa G, Jones L, Tripp R, Zhao Y. An Au/Si hetero-nanorod-based biosensor for Salmonella detection. Nanotechnology 2008;15:e155502.

153. Hahn MA, Keng PC, Krauss TD. Flow cytometric analysis to detect pathogens in bacterial cell mixtures using semiconductor quantum dots. Anal Chem 2008;3:864e72.

154. Yang L, Li Y. Quantum dots as fluorescent labels for quantitative detection of S. typhimurium in chicken carcass wash water. J Food Protect 2005;6:1241e5.

155. Tully E, Hearty S, Leonard P, O'Kennedy R. The development of

rapid fluorescence-based immunoassays, using quantum dot-labelled antibodies for the detection of L. monocytogenes cell surface proteins. Int J Biol Macromol 2006;1e3:127e34.

156. Wang C, Irudayaraj J. Gold nanorod probes for the detection of multiple pathogens. Small 2008;12:2204e8.
157. Su SL, Li Y. Quantum dot biolabeling coupled with immunomagnetic separation for detection of Escherichia coli O157:H7. Anal Chem 2004;76(16):4806e10.
158. An J, Zhang M, Wang S, Tang J. Physical, chemical and microbiological changes in stored green asparagus spears as affected by coating of silver nanoparticles-PVP. LWT e Food Sci Technol 2008;41(6):1100e7.
159. Fernandez A, Soriano E, Lopez-Carballo G, Picouet P, Lloret E, Gavara R. Preservation of aseptic conditions in absorbent pads by using silver nanotechnology. Food Res Int 2009;42(8):1105e12.
160. Fernandez A, Picouet P, Lloret E. Cellulose-silver nanoparticle hybrid materials to control spoilage-related microflora in absorbent pads located in trays of fresh-cut melon. Int J Food Microbiol 2010a;142(1e2):222e8.
161. Fernandez A, Picouet P, Lloret E. Reduction of the spoilagerelated microflora in absorbent pads by silver nanotechnology during modified atmosphere packaging of beef meat. J Food Protect 2010b;73(12):2263e9.
162. Emamifar A, Kadivar M, Shahedi M, Soleimanian-Zad S. Evaluation of nanocomposite packaging containing Ag and ZnO on shelf life of fresh orange juice. Innovat Food Sci Emerg Technol 2010;11(4):742e8.
163. Emamifar A, Kadivar M, Shahedi M, Soleimanian-Zad S. Effect of nanocomposite packaging containing Ag and ZnO on inactivation of Lactobacillus plantarum in orange juice. Food Control 2011;22(3e4):408e13.
164. Jin T, Gurtler JB. Inactivation of Salmonella in liquid egg albumen by antimicrobial bottle coatings infused with allyl isothiocyanate, nisin and zinc oxide nanoparticles. J Appl Microbiol 2011;110(3):704e12.
165. Li H, Li F, Wang L, Sheng J, Xin Z, Zhao L, et al. Effect of nano-packing on preservation quality of Chinese jujube (Ziziphus jujuba Mill. var. inermis (Bunge) Rehd). Food Chem 2009;114(2):547e52.

166. Zhou L, Lv S, He G, He Q, Shi BI. Effect of PE/AG2O nanopackaging on the quality of apple slices. J Food Qual 2011;34(3):171e6.

167. Vikesland PJ, Wigginton KR. Nanomaterial enabled biosensors for pathogen monitoring e a review. Environ Sci Technol 2010;10:3656e69.

168. Dungchai W, Siangproh W, Chaicumpa W, Tongtawe P, Chailapakul O. S. typhi determination using voltammetric amplification of nanoparticles: a highly sensitive strategy for metalloimmunoassay based on a copper-enhanced gold label. Talanta 2008;2:727e32.

169. Yang L, Li Y. Quantum dot bioconjugates for simultaneous detection of Escherichia coli O157:H7 and Salmonella Typhimurium. Analyst 2006;131:394e401.

170. Zhao X, Hilliard L, Mechery S, Wang Y, Bagwe R, Jin S. A rapid bioassay for single bacterial cell quantitation using bioconjugated nanoparticles. Proc Natl Acad Sci USA 2004;42:15027e32.

INDEX

L

M

N

O

P

Q

R

S